Formulas for the E6-B Air Navigation Computer

Using the E6-B Simply & Efficiently

By Frank E. Hitchens

Andrews UK Limited

First published worldwide by
Andrews UK Limited
The Hat Factory
Bute Street
Luton, LU1 2EY

www.andrewsuk.com

Contents

8 – Centre of Gravity

9 – Wind Slide Computer

Introduction

Introduction to this book

This book *Formulas for the E6-B Air Navigation Computer* has been written with the Private, Commercial and Air Line Transport Pilot in mind. A complete set of formulas for the E6-B air navigation computer are included here all in one volume, covering the commonly used time, speed & distance problems, air speed and altitude conversions, fuel, centre of gravity, Pressure Pattern Navigation, and en route navigation problems, and more. On the wind-slide computer, all calculations required for true heading & true air speed, track & ground speed and wind velocity are included plus wind finding methods, etc, in fact, all a pilot needs to know in one book.

However, in my original manuscript first typed up many years ago, I had included Pressure Pattern Navigation, a requirement for the ALTP licence in those days. With the advent of GPS and also Inertial Navigation Systems in many larger aircraft, Pressure Pattern Navigation became virtually redundant. Therefore, I agonised over removing it from this final manuscript; however, my final decision was to keep it in, in the belief someone, some day, somewhere, may find it interesting and of practical use.

To the right of each formula title can be found inserted one of the following sets of letters or a symbol where:

PPL = Private Pilot Licence
CPL = Commercial Pilot Licence
ATPL = Airline Transport Pilot Licence
NAV = Flight Navigator Licence
* = 2 in 1 formula, or two-shot method

The letters indicate the pilot or navigator licence level at which the formula is associated with. Therefore, a student pilot would study the PPL formulas first before moving on to the CPL and higher licence level formulas. The titles labelled as NAV are associated with the Pressure Pattern Navigation formulas required by flight navigators. The asterisk (*) indicates a formula that will give two answers in one formula (two-shot method); it should be used in preference to save time, instead of working two separate associated formulas.

Having bought your own E6-B with its own brief instructional booklet, you should be familiar with some of the basic formulas, which have also been included in this book for completeness. As you work through this book and come across new formulas, and methods as in Section 9 – Wind Triangle Computer, try doing self-tests of each example by inserting your own figures for practice to ensure you are thoroughly familiar on how to perform the calculations.

Some problems will be familiar but presented with an alternative method. Other problems can be performed with the two-shot method mentioned above, where one adjustment of the slide rule can solve two related problems in one setting. All numbers used in the examples have been chosen at random and do not represent any particular aircraft make, model or power plant – they are purely for practice purposes only.

The most common units for air navigation are the nautical mile for distance, Knots for speed, feet for altitude and temperature in degrees Celsius; these units are used mostly within this book. However, fuel is still worked in Imperial Gallons, US Gallons and Litres depending on your location in the world. Caution is required when using one system of units for fuel and then converting to another system.

- Hint: Remove the compass rose and wind-slide and write any formulas you have difficulty remembering on the blank circle area covered by the slide. It may come in handy one day!
- Hint: On the circular slide rule, place the cursor over the required number on the moveable inner scale first, then rotate the inner scale to align the cursor over the appropriate number on the outer scale.
- Hint: A handy item is a thumb rubber: this is a small rubber ring that fits over the end of your thumb to help grip the rotating circular slide rule. It does not interfere with flying the plane, so it can be worn throughout the flight.

Introduction to the E6-B Circular Slide Rule
The straight slide rule was invented c1620–30 and improved upon in 1630 by William Oughtred (1576–1660) an Anglican minister, who also invented the first circular slide rule. The slide rule is a mechanical analogue device used originally for solving complex calculus computations. Prior to WW II, the American, Lieutenant Philip Dalton, introduced the first circular slide rules for use by military pilots. It was first known as the Dalton Dead Reckoning Computer and with further refining it became known by various model designations, until 1940 when it was given a military designation and a patent as the E-6B. The circular slide was later made by other companies after the patent period expired and was given similar designations such as, E6B or E6-B where the hyphen was removed or changed position. Other companies devised their own titles: the English company Airtour, for example, used the series CRP-1 through CRP-5, etc. Today, pilots throughout

the world refer to them colloquially as slipsticks, prayer wheels or whiz wheels, or more correctly as a flight computer. However, throughout this book I will refer to the circular slide rule as the E6-B.

In 1972, the HP-35 electronic calculator was introduced replacing the common slide rules, which are now collectors' items. In more recent years, glass panel cockpits, GPS units, iPads and Electronic Flight Bags (EFB) etc, have become common as pilot accessories, performing some of the functions of the E6-B making it almost redundant. However, any electrical device, or glass panel can fail; a plastic or metal E6-B is the best back-up device being fail-proof, as long as you know how to use it and practice solving problems frequently. Therefore, you as a pilot should be proficient in the use of your E6-B computer at all times. It is also required for use in some pilot exams, therefore knowledge and the use of the circular slide rule is essential.

The modern E6-B's of today come in two different sizes, the small, shirt-pocket size and the larger version, which corresponds with the traditional ten-inch straight slide rules and is more accurate than the smaller type.

The inner and outer scales on the slide rule side are based on common logarithms; other scales are based on mathematical constants used for conversion factors. Air speed and altitude and density altitude windows can be placed anywhere as long as the fixed and rotating scales are aligned to give correct results in association with the inner and outer scale readings. Other scales include some or all of the following: Imp G, US G, Ltrs, Feet, Yards, Metre, NAUT. M., SM, Km, a Latitude scale for Pressure Pattern Navigation and also a Compressibility window, etc, which are placed variously around the circular slide rule dependant on their conversion factors. English made E6-B's have additional scales for conversions between Imperial and Metric units, which are not always found on US made E6-B's. The original Dalton E6-B's did not have all these scales, which have been added to modern E6-B's over the years as they became relevant. Likewise, the original wind-slide scale only covered the relatively lower air speeds of WW II aircraft. With the advent of high-speed jet aircraft, the wind-slide was extended to cover speeds up to around 1000 Knots (or MPH or KMH).

Some air navigation computers are sold with a rotating wind speed scale affixed to the grommet on the wind-slide side of the computer. This can be a handy feature for the Private Pilot – it negates the need to draw pencil lines or a wind-cross. However, when moving on to more complex problems, triple drift calculations for example, the rotating slide gets in the way of drawing pencil lines. I removed the one on my computer as soon as I bought it.

The first E6-B computer this author bought in 1966 was a metal type, which I soon lost two years later in a shipwreck, no less! It was replaced with a CRP-5 plastic type made by Airtour in England, which gave me good service until I left it in the cockpit of a plane and it warped in the heat of the sun. This too, was replaced with another of the same make and type and I still have it after nearly forty years. Having used both metal and plastic, I found the metal type can be rough to rotate with metal fillings underneath the disc, as opposed to the plastic type, which are smoother to operate and a far better product, in this author's opinion. Each to his own, use your preference.

Fly safely with knowledge.

Frank Hitchens,
Wellington, NZ.

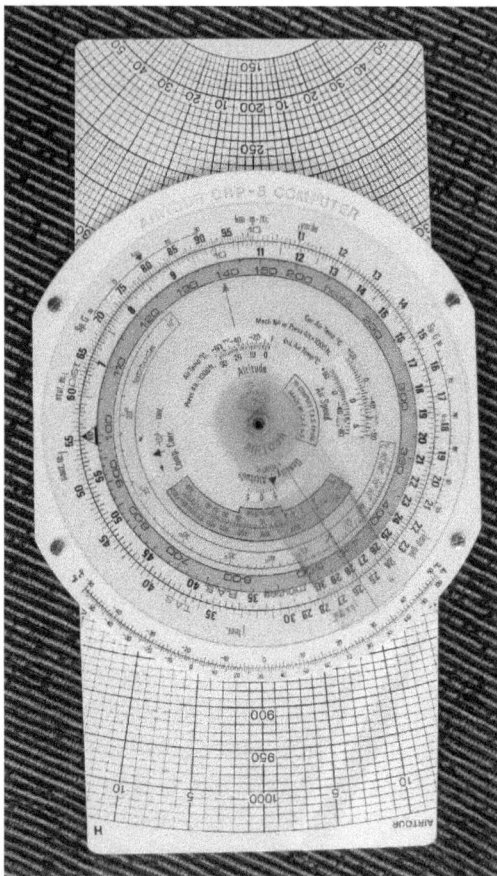

The circular slide rule

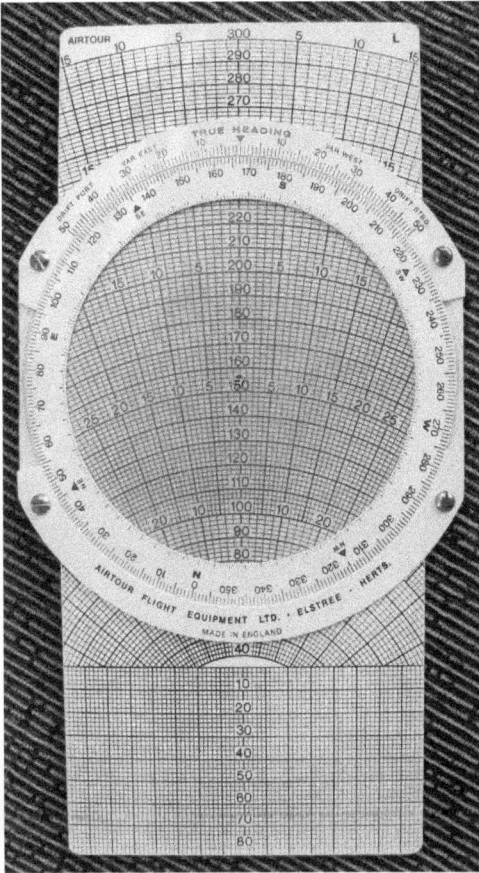

The wind-slide computer

1 – Mathematics

Introduction

In this section, the E6-B air navigation computer slide rule is used to solve various maths problems as an introduction to the formulas appearing from section two onwards. It is assumed the reader has an understanding of basic mathematical procedures, which form the basis for the E6-B calculator slide rule. The aim here is to be able to use the E6-B confidently to solve any navigation problems presented.

1 – Multiplication

Set the '10' index on the inner scale to the multiplier on the outer scale. Search for the multiplicand on the inner scale, and find the product, answer above on the outer scale. Insert the decimal point where appropriate. Always make a mental estimate of all answer before accepting the slide rule answer.

Formula:

$$\frac{\text{Outer scale}}{\text{Inner scale}} \cdots \frac{\text{Multiplier}}{10} \cdots \frac{\text{Product}}{\text{Multiplicand}} = \text{Answer}$$

Given: 5 times 15

$$\frac{5}{10} \cdots \frac{75}{15} = \text{Answer}$$

2 – Division

Division is easily accomplished on the E6-B and forms the basis of several formulas. Set divisor on inner scale below dividend on the outer scale, read the quotient, answer above the '10' index on the outer scale. Cross multiply to check answer after rounding up or down for an estimate answer.

Formula:

$$\frac{\text{Outer scale}}{\text{Inner scale}} \cdots \frac{\text{Dividend}}{\text{Divisor}} \cdots \frac{\text{Quotient}}{10} = \text{Answer}$$

Given: 25 ÷ 5 = ?

$$\frac{25}{5} \cdots \frac{5}{10} = \text{Answer}$$

3 – Multiplication & Division

Presented with a combination of multiplication and division, start by dividing and then multiplying, alternating continuously until all numbers has been solved.

Formula:

$$\frac{\text{Outer scale}}{\text{Inner scale}} \dots \frac{x}{10} \dots \frac{y}{z} = \text{Answer}$$

Given:

$$\frac{7 \times 6}{4 \times 2} = \text{Answer}$$

$$\frac{7}{4} \dots \frac{1.75}{10} \dots \frac{10.5}{6} \dots \frac{10.5}{2} = \frac{5.25}{10} = \text{Answer}$$

4 – Fractions to Decimals

Conversion between fractions, decimals and percentages is quite simply on the E6-B slide rule. Set fraction denominator on the inner scale below the numerator on the outer scale. Over the '10' index, read the decimal equivalent. Multiply the decimal by 100 to obtain the percentage.

Formula:

$$\frac{\text{Outer scale}}{\text{Inner scale}} \cdots \frac{\text{Numerator}}{\text{Denominator}} \cdots \frac{\text{Decimal or \%}}{10} = \text{Answer}$$

Given: ¼ = 0.25 or 25%

$$\frac{1}{4} \cdots \frac{25}{10} = \text{Answer}$$

5 – Percentage of a Given Number

Given a number, the percentage can be found by setting the percentage value on the outer scale over the '10' index. Move to the given number on the inner scale and find the answer above it on the outer scale.

Formula:

$$\frac{\text{Outer scale}}{\text{Inner scale}} \cdots \frac{\text{Given \%}}{10} \cdots \frac{\text{Find value}}{\text{Given number}} = \text{Answer}$$

Given: 25% of 400

$$\frac{25\%}{10} \cdots \frac{100}{400} = \text{Answer}$$

6 – Ratios

Ratios are used to compare one quantity with another quantity in the same units. Ratios are written with the word 'to' or equal sign or as a fraction. Cross multiplying both ways will produce two answers the same if the ratio has been correctly calculated.

Formula:

$$\frac{\text{Outer scale}}{\text{Inner scale}} \ldots \frac{\text{Numerator}}{\text{Denominator}} \ldots \frac{\text{Numerator}}{\text{Denominator}} = \text{Answer}$$

Given: $8 : 16 :: x : 6 = \text{Answer}$

$$\frac{8}{16} \ldots \frac{3}{6} = \text{Answer}$$

Note: $6 \times 8 = 48$ and $3 \times 16 = 48$. Therefore, the answer is correct.

7 – Rates

A rate compares two quantities with two different units. Time, speed and distance, or fuel consumption, etc, all use rate problems, e.g. distance per time, or miles per gallon.

Example 1:

Formula:

$$\frac{\text{Outer scale}}{\text{Inner scale}} \cdots \frac{\text{Unit 1}}{\text{Unit 2}} \cdots \frac{\text{Rate}}{10} = \text{Answer}$$

Given: Total pay of $450 for 30 hours work; find the hourly rate of pay.

$$\frac{\$450}{30} \cdots \frac{\$15 \text{ per hour}}{10} = \text{Answer}$$

Example 2:

Formula:

$$\frac{\text{Rate}}{10} \cdots \frac{\text{Unit 1}}{\text{Unit 2}} = \text{Answer}$$

Given: Find total pay for working 45 hours at $18 per hour.

$$\frac{\$18 \text{ per hour}}{10} \cdots \frac{\$810}{45 \text{ hours}} = \text{Answer}$$

Example 3:

Given: How long must a man work to earn $650 at a rate of pay of $21.00 per hour?

$$\frac{\$650}{\$21} \cdots \frac{31 \text{ hours}}{10} = \text{Answer}$$

8 – Reciprocals

The reciprocal of a given number is written as $1/x$ (one over x) or 'Y' divided by the given number.

Formula:

$$\frac{\text{Outer scale}}{\text{Inner scale}} \cdots \frac{\text{Numerator}}{\text{Denominator}} \cdots \frac{\text{Quotient}}{10} = \text{Answer}$$

Given: Reciprocal of 5

$$\frac{10}{5} \cdots \frac{0.2}{10} = \text{Answer}$$

2 – Navigation Problems

Introduction

In the first section of this book, we looked at the basic mathematical principles as applied to the slide rule. Moving on to this section on navigation problems, we can now put these principles into practice, and look at different ways to use the E6-B computer slide rule to solve problems that occur in day-to-day flight planning and practical air navigation.

We start with one of the most used formulas, the 'Time, Speed & Distance' formula before moving on to other associated formulas, following on with simple Track Intercepts and the Equi-time Point formula. Conversions using the Air Speed and Altitude windows comes next, followed by formulas for vertical navigation (VNAV), and the '1 in 60' rule.

This must be one of the most common formulas of all, and it is included in this book for completeness. Given any two factors, the third can be easily found.

Formula:

$$\frac{\text{Speed}}{60 \text{ Rate Index}} \cdots \frac{\text{Distance}}{\text{Time}} = \text{Answer}$$

Example 1: Find time.

Given:
Speed   = 125 Knots
Distance = 85 NM

$$\frac{125 \text{ Knots}}{60} \cdots \frac{85 \text{ NM}}{42 \text{ Minutes}} = \text{Answer}$$

Note: The answer for time is on the inner scale.

Example: 2: Find speed.
Given:
Time   = 18 minutes
Distance = 72 NM

$$\frac{72 \text{ NM}}{42 \text{ Minutes}} \cdots \frac{240 \text{ Knots}}{60} = \text{Answer}$$

Note: The answer for speed is on the outer scale.

Example 3: Find distance.

Given:
Time   = 11 minutes
Speed   = 142 Knots

$$\frac{142 \text{ Kts}}{60} \cdots \frac{72 \text{ NM}}{11 \text{ Minutes}} = \text{Answer}$$

Note: The answer for distance is on the outer scale.

10 – Instrument Approach Timing

This is a time, speed and distance formula for use on short distances, such as on instrument approaches for timing in minutes and seconds between the outer marker, inner marker and the decision height or missed approach point.

Use the '36' index (for 3600 seconds/hour) as the reference figure in lieu of the '60' Rate index. The hours scale is then read as minutes and the minutes scale is read as seconds. A pencil mark can be made at the '36' position for easy reference.

Formula:

$$\frac{\text{Ground Speed}}{\text{36 Index}} \cdots \frac{\text{Distance}}{\text{Time (seconds)}} = \text{Answer}$$

Given:
Speed   = 95 Knots
Distance = 4.5 NM

$$\frac{\text{95 Knots}}{\text{36}} \cdots \frac{\text{4.5 NM}}{\text{170 seconds (2m 50 secs)}} = \text{Answer}$$

11 – Decimal Time

Decimal time may be used for endurance time in aircraft performance charts, Pilots log books, and Hobbs meters, etc. To convert between decimal time and common clock time, set the '10' index on the inner scale below '60' on the outer scale. Read off the decimal time on the inner scale below clock time or vice versa.

Formula:

$$\frac{60}{10} \cdots \frac{\text{Clock time}}{\text{Decimal times}} = \text{Answer}$$

Given:
Decimal time = 0.75 minutes

$$\frac{60}{10} \cdots \frac{45 \text{ minutes}}{0.75} = \text{Answer}$$

12 – Percentage of Time

Flight planning for larger aircraft may require a percentage of the time en route to be added to the flight plan's total trip time.

Formula:

$$\frac{\text{En Route Time}}{10} \dots \frac{\text{Additional Trip Time}}{\text{Reserve \%}} = \text{Answer}$$

Given:
Trip time   = 300 minutes
Reserve % = 2%

$$\frac{300 \text{ Minutes}}{10} \dots \frac{6 \text{ Minutes}}{2\%} = \text{Answer}$$

13 – Ground Speed in Nautical Miles per Minute

To work some formulas requires the ground speed to be converted from Knots (or MPH) to nautical miles per minutes (NM/min). The answer is found over the '10' index, where 10 represents one minute.

Formula:

$$\frac{\text{Ground Speed (Knots)}}{60 \text{ Rate}} \cdots \frac{\text{NM/minute}}{10} = \text{Answer}$$

Given:
Ground speed = 180 knots

$$\frac{180 \text{ Knots}}{60} \cdots \frac{3 \text{ NM/minute}}{10} = \text{Answer}$$

14 – Conversion of Speed & Distance

This formula makes conversions between nautical miles, statute miles and kilometres, (or Knots, MPH and KPH), which use the same index marks. The index marks are located above the outer scale.

Example:

Speed conversions:

Index:	Naut. M	Stat. M	Km – M – Ltr
Speed:	156 Knots	180 MPH	289 KMH

Distance conversions:

Index:	Naut. M	Stat. M	Km – M – Ltr
Distance:	65 NAUT. M.	75 SM	121 Km

Index:	Km – M – Ltr	Yards	Feet
Distance:	145 metres	158 yards	475 feet

Note: Not all E6-B's carry the index marks for feet, yards and metres. These index marks can be placed on your computer if required at the following places on the outer scale:

- Above 10 mark 'Km-Metres-Litres'

- Above 10.9 on the outer scale mark 'Yards'

- Above 32.8 mark 'Feet'.

See Conversion factors.

15 – Time and Distance to Navigation Aid   * IR

The time and distance to a nav aid can be found using this two-shot method on the E6-B computer.

Formula:

$$\frac{\text{Minutes Flown}}{\text{Bearing Change}} \cdots \frac{\text{Time to Nav Aid}}{60} \cdots \frac{\text{Distance to Nav Aid}}{\text{TAS}} = \text{Answer}$$

Given:
Minutes flown   = 5 minutes
Bearing change = 10°
True Air Speed = 140 KTAS

$$\frac{\text{5 Minutes}}{10°} \cdots \frac{\text{30 Minutes}}{60} \cdots \frac{\text{70 NM}}{140 \text{ TAS}} = \text{Answer}$$

16 – Off-track DME Distance

If a DME is located up to 30 degrees off-track, the DME distance read-out can be converted to a true distance, to a position on track abeam the DME beacon. (See also, Formula 89, DME Off-track Corrected Reading).

Formula:

$$\frac{\text{DME Distance}}{10} \cdots \frac{\text{True Distance NM}}{\text{Cosine Factor}} = \text{Answer}$$

Given:
DME read-out= 65 NM
Cosine 25°   = 0.90

$$\frac{65}{10} \cdots \frac{58.5}{90} = \text{Answer}$$

Cosine Table
5° = 0.99
10° = 0.98
15° = 0.96
20° = 0.94
25° = 0.90
30° = 0.86

17 – Interception, Same Track & Direction   ATPL

When two aircraft are on the same track and flying in the same direction but at different speeds, the time and distance that the faster aircraft overtakes the slower aircraft can be calculated using the 'speed of closing'. The speed of closing is the difference of the two aircraft's ground speeds, and it is used in the same way as the speed for normal time, speed and distance problems.

Formula:

$$\frac{\text{Speed of Closing}}{60} \cdots \frac{\text{Distance}}{\text{Time}} = \text{Answer}$$

Find the time aircraft 'A' passes aircraft 'B' and the distance from position 'Y' on track.

Given:
Aircraft 'A' G/S   = 270 Kts
Aircraft 'B' G/S   = <u>-220 Kts</u>
Speed of Closing =   50 Kts
Separation   = 42 NM
Aircraft 'A' passes 'Y' = 11.15 AM

Interception time is:

$$\frac{50 \text{ Kts}}{60} \cdots \frac{42 \text{ NM}}{50 \text{ Minutes}} = \text{Answer}$$

The time aircraft 'A' overtakes aircraft 'B' is 11.15 + 00.50 = 12.05 PM = *Answer A*

Distance of intercept from position 'Y' = 50 minutes at 270 Kts = 225 NM = *Answer B*

Note: The answer is on the inner scale.

18 – Interception, Opposing Heading

To find the time and distance two aircraft will pass head-on, on the same track, the ground speed of each aircraft is added together to find the speed of closing.

Formula:

$$\frac{\text{Speed of Closing}}{60} \cdots \frac{\text{Distance}}{\text{Time}} = \text{Answer}$$

Given:
Aircraft 'A' departs from 'X' at = 250 Kts
Aircraft 'B' departs from 'Y' at = +175 Kts
Speed of Closing   = 425 Kts
'X' and 'Y' are 200 NM apart

Find the time and distance from 'X' when the two aircraft will pass.

A) Find interception time is 200 NM at 425 Kts

$$\frac{425 \text{ Kts}}{60} \cdots \frac{200 \text{ NM}}{28 \text{ Minutes}} = \text{Answer}$$

Note: The answer is on the inner scale.

B) Find distance 28 minutes at 250 Kts

$$\frac{250 \text{ Kts}}{60} \cdots \frac{116 \text{ NM}}{28 \text{ Minutes}} = \text{Answer}$$

19 – Radius of Action

On first inspection of this formula it appears contradictory to find 'ground speed out' over 'time home' and vice versa. The reason for this anomaly is the fact, inverse proportions are used here, and remember, time required to cover a given distance is inversely proportional to speed. The radius of action involves flying the same distance 'out' and 'home' and the ratio of the time out and back is proportional to ground speed out and back.

Set 'G/S out plus G/S home' over fuel endurance. Find 'G/S out' on the outer scale and read 'time home' on the inner scale. Repeat for 'G/S home' to find 'time out'. 'Time out and home' must equal safe fuel endurance.

Note: This formula is a two-shot method.

Formula:

$$\frac{\text{G/S Out} + \text{G/S Home}}{\text{Fuel Endurance}} \cdots \frac{\text{G/S Out}}{\text{Time Home}} \cdots \frac{\text{G/S Home}}{\text{Time Out}} = \text{Answer}$$

Given:
Fuel endurance = 04.15 = 255 minutes
G/S out   =   150 Kts
G/S home   = +175 Kts
G/S out + home =   325 Kts

Find: Time out and home:

$$\frac{325 \text{ Kts}}{255 \text{ Minutes}} \cdots \frac{150 \text{ Kts}}{01.58} \cdots \frac{175 \text{ Kts}}{02.18} = \text{Answer}$$

Confirm time: 01.58 plus 02.18 = 04.15.
Note: The answer is on the inner scale.

20 – Equal Time Point (Critical Point)   CPL

The equal-time (or equi-time point) is the point on track where it will take as long to fly to the destination as it will to return to the departure point. It is also known as the 'critical point'. Ground speed out may also be called 'ground speed on'. The TAS out and home is assumed to remain constant in this formula.

Formula:

$$\frac{\text{G/S Out} + \text{G/S Home}}{\text{G/S Home}} \cdots \frac{\text{Total Distance}}{\text{ETP Distance}} = \text{Answer}$$

Given:

Total distance	=	620 NM
G/S out	=	195 Kts
G/S home	=	+210 Kts
G/S out + home	=	405 Kts

$$\frac{405 \text{ Kts}}{210 \text{ Kts}} \cdots \frac{620 \text{ Kts}}{322 \text{ NM}} = \text{Answer (ETP Distance)}$$

Note: The answer is on the inner scale.

21 – ETP³ & ETPᴰ

Flight planning for multi-engine aircraft usually requires the calculation of an equal-time point allowing for reduced air speed in the event of an engine failure or due to depressurized flight at lower altitude. The terms ETP³ is used for flight on three engines and ETPᴰ is used to denote depressurized flight. An engine failure or depressurized flight will result in a lower TAS. Of course, the formula applies for any multi-engine aircraft.

Formula:

$$\frac{\text{Reduced G/S Out} + \text{Reduced G/S Home}}{\text{Reduced G/S Home}} \cdots \frac{\text{Distance}}{\text{ETP}} = \text{Answer}$$

Given:

Reduced G/S out	=	252 Kts
Reduced G/S home	=	+286 Kts (to Inner scale)
G/S out + G/S home	=	538 Kts (to Outer scale)
Normal G/S (cruise)	=	310 Kts
Total distance	=	950 NM

$$\frac{538 \text{ Kts}}{286 \text{ Kts}} \cdots \frac{950 \text{ NM}}{505 \text{ NM}} = \text{Answer (ETP Distance)}$$

Note: The reduced G/S home and reduced G/S out plus G/S home are reversed positions on the inner and outer scale. Time to ETP is flown at normal cruise G/S. From ETP, time to continue to destination at reduced G/S out is the same as the time to return at reduced G/S home. The answer is on the inner scale.

24

22 – Confirming ETP Distance

This formula confirms the ETP[3] distance and times to return from the ETP and time ETP to destination.

After calculating the distance to the ETP in the previous formula, the figures can be confirmed by setting the ground speed out on the outer scale over the ground speed home on the inner scale. On the outer scale read the distance from the ETP to the destination (out) over the distance on the inner scale, from the ETP back to the departure point (home). The distance added together should equal and confirm the total distance. Calculating the speed and distance out and return should confirm the equal times.

Formula:

$$\frac{\text{G/S Out}}{\text{G/S Home}} \cdots \frac{\text{Distance ETP Out}}{\text{Distance ETP Home}} = \text{Answer}$$

Given:
Reduced G/S out   = 252 Kts
Reduced G/S home   = 286 Kts
Total distance   = 950 NM

Confirmation 1:

$$\frac{252 \text{ Kts}}{286 \text{ Kts}} \cdots \frac{445 \text{ NM}}{505 \text{ NM}} = \text{Answer}$$

Confirmation 2:
Return at ETP[3] at 286 Kts over 505 NM = 01.46 hours
Continue from ETP[3] at 252 Kts over 445 NM = 01.46 hours

23 – Circle of Probability

The circle of probability is the area contained within a given radius centred on the DR position, in which the aircraft has a 50% chance of being in. The radius depends on the aircraft's speed range and time between the last fix and the DR position. It is based on the following:

Aircraft's Speed Range	1 Hour Radius
120–220 Kts	8 NM
220–430 Kts	14 NM
430–560 Kts	20 NM

Formula:

$$\frac{1 \text{ Hour Radius}}{60} \dots \frac{\text{Circle of Probability}}{\text{Time Interval}} = \text{Answer}$$

Given:
1 Hour radius = 14 NM
Time interval = 40 minutes
Ground speed = 285 Kts

$$\frac{14 \text{ NM}}{60} \dots \frac{9.3 \text{ NM}}{40 \text{ Minutes}} = \text{Answer}$$

24 – Cross-wind Component & Drift

The in-flight cross-wind component can be found (on the outer scale) given the ground speed and drift. Alternatively, given ground speed and cross-wind component, the drift can be found (on the inner scale).

Formula:

$$\frac{\text{Ground Speed}}{60} \cdots \frac{\text{Cross-wind Component}}{\text{Drift}} = \text{Answer}$$

Given (A):
Ground speed   = 165 Kts
Drift   = 12°

$$\frac{165 \text{ Kts}}{60} \cdots \frac{33 \text{ Kts}}{12°} = \text{Answer}$$

Note: The answer is on the outer scale.

Given (B):
Ground speed   = 180 Kts
Cross-wind component = 25 Kts

$$\frac{180 \text{ Kts}}{60} \cdots \frac{25 \text{ Kts}}{8.35°} = \text{Answer}$$

Note: The answer is on the inner scale.

25 – Weighted Wind Component

When flight-planning long distance flights, knowledge of the average wind component is required. This can normally be found by averaging all the wind components for each stage of the flight. However, when the stages are of different length, the wind components will need to be 'weighted' before they can be averaged out in the normal way.

Formula:

$$\frac{\text{Stage Time}}{\text{Total Trip Time}} \cdots \frac{\text{Weighted Wind Component}}{\text{Head or Tail-wind Component}} = \text{Answer}$$

Given:
Total trip time   = 285 minutes
Stage time   = 65 minutes
Head or Tail-wind Component = −35 Knots

$$\frac{65 \text{ minutes}}{285 \text{ Minutes}} \cdots \frac{-7 \text{ Knots}}{-35 \text{ Knots}} = \text{Answer}$$

3 – Air Speed

Air Speed Definitions

The air speed indicator is calibrated to 1013.25 hPa (29.92 inches) at 15° Centigrade at sea-level and for normal aircraft flight attitudes. Any variation from these ambient factors or aircraft attitude will produce erroneous readings and must be allowed for during navigation calculations involving air speed.

- IAS = indicated air speed, as read from the ASI
- IAS plus instrument error and position/pressure error (IE and PE) = Calibrated air speed (CAS) or Rectified air speed (RAS). CAS is used for aircraft performance
- CAS (RAS) minus compressibility error = Equivalent air speed (EAS). It depends on pressure altitude and relative air density
- EAS plus or minus temperature correction and density error = True air speed (TAS)
- TAS is the speed of the aircraft relative to the surrounding air, and is used for navigational purposes.

Note: IAS + IE & PE = CAS

- For light aircraft use IAS (CAS) plus or minus temperature and density error to find TAS. Compressibility is ignored.
- For flight above 200 Knots and 10 000 feet, convert CAS to EAS allowing for compressibility and then convert from EAS allowing for plus or minus temperature and density error to find TAS.
- The term CAS is favoured by the Americans, and the term RAS is favoured by the English. American and English made E6-B's will be marked respectively in CAS or RAS.

Compressibility Correction Tables

Above 200 Kts and 10 000 feet altitude, the calibrated air speed (CAS) has to be corrected for compressibility effects to find the equivalent air speed (EAS).

Two different methods are presented here:

- Table 'A' multiplication method: Multiply the CAS by the 'F' factor to find the equivalent air speed (EAS).
- Table 'B' subtraction method: Subtract the factor from the CAS to find the EAS. This table ends at the Mach 1.0 limit.

Table 'A' (Multiplication Method)

| | | | | CAS Kts | | | |
| | 200 | 250 | 300 | 350 | 400 | 450 | 500 |
Pressure Altitude				'F' Factor			
10 000 ft.	1.0	1.0	.99	.99	.98	.98	.07
20 000 ft.	.99	.98	.97	.97	.96	.95	.94
30 000 ft.	.97	.96	.95	.94	.92	.91	.90
40 000 ft.	.96	.94	.92	.90	.88	.87	.86
50 000 ft.	.93	.90	.87	.86	.84	.84	.84

Table 'B' (Subtraction Method)

| | | | | | CAS Kts | | | | | |
| | 150 | 200 | 250 | 300 | 350 | 400 | 450 | 500 | 550 | 600 |
Pressure Altitude					Subtraction Factor					
5 000 ft.	0	0	1	2	2	3	5	6	8	10
10 000 ft.	0	1	2	3	5	7	10	13	17	21
15 000 ft.	1	2	3	5	8	12	16	21	27	
20 000 ft.	1	3	5	8	12	17	23	31		
25 000 ft.	2	4	7	11	17	24	32			
30 000 ft.	2	5	9	15	23	32				
35 000 ft.	3	7	12	20	29					
40 000 ft.	4	9	16	25						

26 – IAS/CAS to TAS

For relatively low-speed and low altitude flight in light aircraft, use this formula. Use the Air Speed Window to correct the indicated (calibrated) air speed (IAS or CAS) for altitude and temperature/density error to find the TAS. The correction may be plus or minus, depending on the air temperature/density error; however, most times it will be plus giving a higher TAS. With standard temperature at altitude CAS (RAS) is always less than TAS due to reducing air density. As a 'rule of thumb', TAS increases by approximately 2 Kts/1000 feet. Therefore, TAS = EAS times the √relative air density.

Method:

1. Use the Air Speed Window to set pressure altitude, (1013 hPa, QFE) and ambient temperature. Indicated altitude (QNH) can be used for practical purposes.

2. Opposite CAS on the inner scale, read TAS on the outer scale.

Example:
Given:
Altitude   = 8000 feet
Temperature = +5°C
CAS   = 135 Kts

Find:
TAS = 154 Kts

27 – Calibrated Air Speed to Equivalent Air Speed

This formula is used for relatively high-speed aircraft, flying at 200 Knots or more at altitudes above 10 000 feet and allowing for compressibility effects on the air speed indicator.

Using the 'F' factor from the compressibility table and the following formula given below, the equivalent air speed (EAS) can be found, given the calibrated air speed and the altitude. The CAS is multiplied by the 'F' factor to find the equivalent air speed (EAS). The next step is to correct EAS for temperature and density error to find TAS. The corrections for temperature, altitude and compressibility will then be completed.

Formula: $\dfrac{CAS}{10} \dots \dfrac{EAS}{\text{'F' Factor}} = $ Answer

Given:
Altitude = 20 000 feet
CAS   = 305 Kts
'F' factor = 0.97

$\dfrac{CAS\ 305\ Kts}{10} \dots \dfrac{EAS\ 296\ Kts}{0.97} = $ Answer

28 – Mach Number from TAS & COAT

Finding the Mach number given the TAS and corrected outside air temperature (COAT). The Mach number index is found inside the Air Speed Window. Some computers do not have a Mach number Index; if this is the case with your E6-B, you can mark in your own Index mark. Set the '10 Index' on the inner scale against 661 Knots on the outer scale. In the Air Speed Window make a mark opposite the +15°C COAT. True Air Speed to Mach number conversions can then be done in the usual manner.

- Note: the speed of sound varies in proportion to the square root of the Absolute temperature (K).
- Note: Conversions from Mach number always gives air speed in Knots only.

1. In the Air Speed Window, set the Mach number Index to the COAT.
2. Opposite TAS on the outer scale, read the Mach number on the inner scale.
3. Read the speed of sound (Mach 1.0) opposite the '10 Index' on the outer scale.

Example:
Given:
COAT   = −34°C
TAS   = 488 Kts

Find: Mach № = 0.81 = *Answer 1*
Speed of sound = 602 Kts = *Answer 2*

29 – Temperature Rise

All E6-B's have a Temperature Rise scale for finding the increase in temperature due to kinetic heating caused by skin friction and compression of the air flow over the aircraft. The kinetic heating causes the outside air temperature gauge to over-read a certain amount depending on the true air speed. The outside air temperature gauge (indicated temperature) always reads higher than true air temperature (TAT). Indicated air temperature minus the temperature rise equals the true air temperature, which is to be used for all calculations involving temperature. The temperature rise scale adjusts the indicated air temperature directly.

If your E6-B has no scale, the temperature rise can be calculated using the following formula and the factors of 950 for air speed in Knots or 1250 for MPH. The TAT is used for air speed and altitude corrections.

Formula:

$$\frac{TAS}{950} \cdots \frac{Temp\ Rise\ °C}{TAS} = Answer$$

Example:
Given:
Factors 950 for Kts & 1250 for MPH
Pressure altitude   = 32 000 feet
Indicated air temp = −45°C
True air temp   = −28°C
CAS   = 242 Kts = TAS 407 Kts
CAS at −28°C TAT = 420 Kts

Use Air Speed Window to find initial TAS 407 Kts and enter into formula:

$$\frac{\text{Trial TAS 407 Kts}}{950} \ldots \frac{17\,°C}{407\,\text{Kts}} = Answer\ 1$$

Indicated air temperature −45°C less +17°C = −28°C TAT = *Answer 2*

- The real TAS can now be found by re-working the problem in the Air Speed Window using the pressure altitude and CAS as given plus the TAT as given by answer 2.
- TAT is also known as corrected outside air temperature (COAT).

4 – Altitude

Altitude Definitions

- Absolute altitude: The altitude above the terrain as indicated by a radar or radio altimeter.
- Pressure altitude: The altitude indicated by the altimeter when set to QFE 1013 hPa (29.92 inches).
- Indicated altitude: The altitude above sea-level when the altimeter is set to the local QNH setting.
- True altitude: The true altitude above sea-level. Indicated and true altitudes are both the same only when standard ISA conditions exist; the altimeter works in the same way as a barometer. True altitude is rarely used.
- Density altitude: Density altitude is pressure altitude corrected for non-standard temperature. It is used for aircraft performance calculations.
- Altitude changes 1°C per 120 feet.
- 1 hPa change = 30 feet, approximately in the lower levels.

30 – True Altitude Temperature Correction   PPL

When flying towards an area of lower temperature the altimeter will indicate a higher altitude than the true altitude, the pressure level being constant. The mnemonic to remember is:

> HALT = Higher (indicated) altitude (because of) lower temperature.

The temperature corrections are seldom used in practice but they can become significant when flying over mountainous areas. A 4–5% difference at sea-level is not much to worry about but at altitude the difference can amount to several hundred feet!

Conversions from indicated altitude to true altitude allowing for temperature changes can be made by setting the indicated altitude and temperature in the Altitude Window. Opposite the Indicated Altitude on the inner scale, find the True Altitude on the outer scale.

Example:
Given:
Temperature   = –15°C
Indicated altitude = 10 000 feet

Find:
True altitude = 9640 feet = Answer

31 – Density Altitude

To find density altitude from temperature and pressure altitude. The performance of an aircraft is based on the density altitude, which quite often is not the same as the true altitude. The density altitude exceeds the pressure altitude by approximately 110 feet/°C above the standard ambient temperature.

Method:
- Use the Air Speed Window.
- Set the temperature to the pressure altitude.
- In the Density Altitude Window, read off the density altitude.

Given:
Temperature   = +25°C
Pressure altitude   = 7000 feet

Find:
Density altitude = 10 000 feet.

5 – VNAV

Introduction

VNAV is the abbreviation for vertical navigation; it involves calculations related to the climb and descent phases of the flight.

32 – Time to Climb or Descend

Calculating the time required on the climb or descent, given the altitude change and the aircraft's rate of climb (RoC).

Formula:

$$\frac{\text{Altitude Change}}{\text{Rate of Climb}} \cdots \frac{\text{Time}}{10} = \text{Answer}$$

Given:

Altitude change = 8 000 feet
Rate of climb (or RoD) = 1450 FPM

$$\frac{8\,000\ \text{Feet}}{1450\ \text{FPM}} \cdots \frac{5.5\ \text{Minutes}}{10} = \text{Answer}$$

33 – Average rate of Climb

The average rate of climb can be determined by timing the climb through two different altitudes.

Formula:

$$\frac{\text{Height Climbed}}{\text{Time}} \dots \frac{\text{Average Rate of Climb}}{10} = \text{Answer}$$

Given: An aircraft climbs from 1000 feet through 3600 feet (a climb of 2600 feet) in 2.5 minutes.

$$\frac{2600 \text{ Feet}}{2.5 \text{ Minutes}} \dots \frac{1040 \text{ FPM}}{10} = \text{Answer}$$

34 – Distance in NM/minute

The next formula (35 – Distance on Climb or Descent) and also other formulas require the ground speed to be converted to NM/minute. Simply divide the ground speed in Knots by 60. Using your E6-B the formula is as follows:

Formula:

$$\frac{\text{Ground Speed Kts}}{60} \ldots \frac{\text{NM/minute}}{10} = \text{Answer}$$

Given:
Ground speed = 240 Kts

Find: NM/minute

$$\frac{240 \text{ Kts}}{60} \ldots \frac{4 \text{ NM/minute}}{10} = \text{Answer}$$

35 – Distance on Climb or Descent CPL

The distance covered during the climb or descent can be found, given the rate of climb or descent, altitude change and ground speed in nautical miles per minute (NM/per minute). To find the NM/minute divide ground speed by 60. See Formulas 34, Distance in NM/minute.

Formula:

$$\frac{\text{Altitude Change}}{\text{RoC/RoD}} \cdots \frac{\text{Distance NM}}{\text{NM/minute}} = \text{Answer}$$

Given:
Altitude change = 8000 feet
Rate of climb   = 900 FPM
Ground speed   = 120 Kts = 2 NM/minute

$$\frac{8000 \text{ Feet}}{900 \text{ FPM}} \cdots \frac{17.8 \text{ NM}}{2 \text{ NM/minute}} = \text{Answer}$$

36 – Time & Distance on Climb or Descent   * CPL

The time and distance covered during the climb or descent can be found in one setting by combining Formula 32 – Time to Climb or Descend, and 34 – Distance in NM/minute. Given altitude change, rate of climb (or descent) and ground speed in NM/minute.

Formula:

$$\frac{\text{Altitude}}{\text{RoC/RoD}} \cdots \frac{\text{Time}}{10} \cdots \frac{\text{Distance}}{\text{NM/minute}} = \text{Answer}$$

Given:
Altitude   = 8500 feet
Roc/Rod   = 750 FPM
Ground speed = 1.5 NM/minute

$$\frac{8500 \text{ Feet}}{750 \text{ FPM}} \cdots \frac{11.3 \text{ Minutes}}{10} \cdots \frac{17 \text{ NM}}{1.5 \text{ NM/minute}} = \text{Answer}$$

37 – Distance on Descent

A comfortable and fuel efficient rate of descent can be achieved by allowing a distance of 5 NM for every 1000 feet to be descended. Multiply the altitude to be descended by 5 (NM) minus the last three numbers, or use the E6-B formula below:

Formula:

$$\frac{5 \text{ NM}}{10} \cdots \frac{\text{ToD Distance}}{\text{Altitude}} = \text{Answer}$$

Given:
Altitude to be lost = 8000 feet

$$\frac{5 \text{ NM}}{10} \cdots \frac{40 \text{ NM}}{8000 \text{ Ft}} = \text{Answer}$$

Note: 5 × 8(000) feet = 40 NM

The rate of descent required depends on the ground speed and increases by 100 FPM for each 30 Knots speed increase.

Ground Speed	Rate of Descent
120 Kts	400 FPM
150 Kts	500 FPM
180 Kts	600 FPM

38 – Climb Gradient (Ft/NM)

The climb gradient may be expressed as feet climbed per nautical mile, with ground speed given in Knots and rate of climb in FPM.

Formula:

$$\frac{\text{Ground Speed}}{60} \cdots \frac{\text{RoC FPM}}{\text{Ft/NM}} = \text{Answer}$$

Note: The answer is found on the inner scale.

Given:
Ground speed = 100 Kts
Rate of Climb = 600 FPM

$$\frac{100 \text{ Kts}}{60} \cdots \frac{600 \text{ FPM}}{360 \text{ Ft/NM}} = \text{Answer}$$

39 – Climb Gradient Percent CPL

Obstacle clearance after take-off may require the aircraft to climb at a minimum percent gradient. Given the ground speed and minimum percentage gradient required, the minimum rate of climb to achieve this climb gradient can be found on the E6-B.

Formula:

$$\frac{\text{Ground Speed Kts}}{10} \dots \frac{\text{RoC FPM}}{\text{\% Gradient}} = \text{Answer}$$

Given:
Ground speed = 120 Kts
% Gradient   = 5%

$$\frac{120 \text{ Kts}}{10} \dots \frac{600 \text{ FPM}}{5\%} = \text{Answer}$$

6 – One-in-Sixty Rule

40 – Off-track Correction (1 in 60)

Given the distance off-track, distance gone, the track error (TE) correction angle to parallel the required track can be found. The problem is re-worked to find the track error to intercept the required track, given distance off-track and distance to go. The sum of the track error to parallel and track error to intercept is added together to find the correction angle, or change in heading, required to reach the destination, if more than half way between the departure point and the destination. Or, having found the track error to parallel (Answer 'A') it can be doubled to arrive back on track the same distance ahead as already gone when the heading change was made.

If your computer has a drift correction window, you may find the answer is less than the answer given over the '60 Rate Index'. The difference is due to the drift correction window being calculated in Radians (57.3° = 1 Radian).

Formula:

$$\frac{\text{Distance Off-track}}{\text{Distance Gone}} \dots \frac{\text{TE to Parallel}}{60} = \text{Answer 'A'}$$

$$\frac{\text{Distance Off-track}}{\text{Distance to Go}} \dots \frac{\text{TE to Intercept}}{60} = \text{Answer 'B'}$$

Given:
Distance off-track = 4 NM
Distance gone   = 30 NM
Distance to go   = 48 NM

$$\frac{4 \text{ NM}}{30 \text{ NM}} \dots \frac{8°}{60} = \text{Answer 'A'} \qquad \frac{4 \text{ NM}}{48 \text{ NM}} \dots \frac{5°}{60} = \text{Answer 'B'}$$

Add answers 'A'+ 'B' (8° + 5°) = 13° = Answer: this is the total heading change required, towards the required track, depending which side of track the aircraft has drifted to.

41 – Diversion

The '1 in 60' rule can be used to find the heading change to an off-track alternate airport. Use the distance from the destination to the alternate airport, as the 'destination to alternate distance' over the 'distance to run' from the point of diversion to destination as 'distance to go'.

Formula:

$$\frac{\text{Destination to Alternate Distance}}{\text{Distance to Run}} \dots \frac{\text{Heading Change}}{60} = \text{Answer}$$

Given:
Destination to alternate distance = 25 NM
Distance to run = 83 NM

$$\frac{25 \text{ NM}}{83 \text{ NM}} \dots \frac{18°}{60} = \text{Answer}$$

42 – Distance Off VOR Radial

The distance between the VOR radial the aircraft is on and the required radial can be found by using the '1 in 60' method. The distance from the VOR beacon is required and the difference between the two radials in degrees, known as 'radial error', is also required. This is basically the same formula as 41 – Diversion, however this time we are looking for the answer in nautical miles, not degrees.

Formula:

$$\frac{\text{Radial error}}{60} \ldots \frac{\text{Distance from Radial}}{\text{Distance from VOR}} = \text{Answer}$$

Given:
Radial error = 9°
Distance from VOR = 47 NM

$$\frac{9°}{60} \ldots \frac{7 \text{ NM}}{47 \text{ NM}} = \text{Answer}$$

7 – Fuel Calculations

Introduction
Fuel calculations are in important part of pre-flight planning in order to determine the required fuel load. During flight, we need to know the quantity of fuel used and fuel remaining for range and endurance calculations. Therefore, fuel calculations fall into three main categories:

- Fuel required, range and endurance, consumption, percentage reserves
- The point of no return (PNR) calculations
- Conversion of units.

Fuel consumption is one of the major expenses in operating an aircraft, and therefore, fuel must be used in the most economical way. For operators of larger commercial aircraft, there is a fine line between carrying sufficient fuel for a flight and tankering excess fuel, which costs money.

Fuel can be measured by volume using Imperial gallons, US gallons or Litres, and also by weight using pounds or kilograms. It is a frequent requirement to convert from one set of units to another, which also brings specific gravity (Sp. G) into the equations.

Depending on the type of flying you are involved in or the navigation examinations you wish to attempt, you may not need to know all the formulas presented in this section. However, after studying those formulas you require, you should have no excuse for exhausting your fuel supply in flight.

Specific Gravity of Fuels Table

Specific gravity (Sp. G) is the ratio of a given volume of fuel to an equal volume of water at 4° Centigrade. The table below shows the specific gravity for some of the more common fuels used by the military and civil operators:

Octane Rating	Specific Gravity
100 LL Avgas	0.72
Jet A, JP1, Avtur	0.79
Jet A1	0.80
JP5 Avcat	0.80
JP4 Avtag	0.757
Synthetic oil	0.96
Mineral oil	0.90

Conversion Factor table

11 Imperial gallons	= 50 Litres	
1 Imperial gallon	= 4.546 Litres	= 1.2 US gallons
1 US Gallon	= 3.785 Litres	= 0.833 Imp. gallons
1 Litre	= 0.22 Imp. gallons	= 0.264 US gallons
1 pound	= 0.45 Kg	
1 Kg	= 2.2 pounds	

43 – Weight Conversions

For conversion between pounds and kilograms. Set the weight in one set of units on the inner scale opposite its index. Read off the equivalent weight on the inner scale opposite its appropriate index. The conversion indexes are located above the Sp. G 80 index for pounds, and the Sp. G 80 index for Kg.

Example:

Index	Kgs	Lbs
Weight	77	170

44 – Volume Conversions

Conversion between Imperial gallons, US gallons and Litres. Set the given quantity on the inner scale opposite its appropriate index *above* the outer scale. Read off the answer on the inner scale opposite the index for the required conversion.

Example:

Index	Litres	Imperial Gallons	US Gallons
Volume	364	80	96

45 – Weight-Volume Conversions

The weight of a given volume of fuel can be found in Pounds or Kilograms by reference to the 'Lbs' or 'Kg' specific gravity scales. The specific gravity scales are located above the outer scale.

Example:
Given:
87 Imperial gallons at Sp. G 0.72, find the weight in Pounds and Kilograms.

Index	Imperial Gallons	Sp. G (Pounds)	Sp. G (Kilograms)
Answer	87	62.6	28.6

46 – Oil Index '90'

The weight of a given volume of oil can be found by setting the '10 Index' on the inner scale opposite the 'Oil Index' (90 or 96) above the outer scale. Read the weight of the oil over the quantity. If there is no 'Oil Index' on your E6-B, set the '10 Index' to 90 (or 96) on the outer scale. Note; the specific gravity scale only goes up to 90, so continue around the outer scale to 96 if calculating for synthetic oil. The weight in pounds or Kg is found in one setting on the E6-B.

Note:
Mineral oil Sp. G   = 0.90
Synthetic oil Sp. G = 0.96

Formula:

$$\frac{\text{Oil Index (90 or 96)}}{10} \ldots \frac{\text{Weight (Lbs)}}{\text{Imp. Gallons}} \ldots \frac{\text{Weight (Kg)}}{\text{Litres}} = \text{Answer}$$

Given:
4 Imperial Gallons oil, find weight.

$$\frac{\text{Oil Index 90}}{10} \ldots \frac{36 \text{ Lbs}}{4 \text{ gallon}} \ldots \frac{16.4 \text{ Kg}}{18.2 \text{ Ltrs}} = \text{Answer}$$

47 – Fuel Required, Consumption & Endurance

By using this one formula, fuel required, consumption and endurance can be found, given any two variables. This formula is similar to the Time, Speed & Distance formula.

Formula:

$$\frac{\text{Consumption}}{60} \cdots \frac{\text{Total Fuel Required}}{\text{Endurance (Time)}} = \text{Answer}$$

Examples:
A – Total fuel required:
Given:
Consumption = 42 Litres/hour
Time = 01.20

$$\frac{42\ \text{Litres/hour}}{60} \cdots \frac{56\ \text{Litres Required}}{01.20} = \text{Answer}$$

B – Fuel consumption:
Given:
Litres consumed = 152 Litres
Time = 04.00

$$\frac{152\ \text{Litres}}{04.00} \cdots \frac{38\ \text{Litres/hour}}{60} = \text{Answer}$$

C – Endurance (Time):
Given:
Consumption = 75 Litres/hour
Total fuel = 275 Litres

$$\frac{75\ \text{Litres/hour}}{60} \cdots \frac{275\ \text{Litres/hour}}{03.40} = \text{Answer}$$

Note: The answer for 'C – Endurance (Time)' is found on the inner scale.

48 – Still Air Range & Endurance   * CPL

Finding the still air range (SAR) and endurance given the total fuel, fuel flow and true air speed. Set total fuel on the outer scale over fuel flow on the inner scale, read off SAR over TAS and endurance in minutes over the '60 Rate Index'. This method finds two answers in one setting.

Formula:

$$\frac{\text{Total Fuel}}{\text{Fuel Flow}} \dots \frac{\text{SAR}}{\text{TAS}} \dots \frac{\text{Endurance}}{60} = \text{Answer}$$

Given:
Total fuel = 390 Pounds
Fuel flow = 90 Lbs/hour
TAS = 175 Kts

$$\frac{390 \text{ Lbs}}{90 \text{ Lbs/hour}} \dots \frac{758 \text{ NM}}{175 \text{ Kts}} \dots \frac{260 \text{ Minutes}}{60} = \text{Answer}$$

Therefore, SAR = 758 NM and Endurance = 260 minutes (04.20)

49 – ETI & Fuel Required   * CPL

When flight planning or during an en route flight, the estimated time interval (ETI) and fuel required to cover a given distance an be found when the distance, ground speed and fuel flow are known. These two problems can be solved in one setting on the E6-B slide rule.

Formula:

$$\frac{\text{Distance NM}}{\text{Ground Speed}} \cdots \frac{\text{ETI}}{60} \cdots \frac{\text{Fuel Required}}{\text{Fuel Flow}} = \text{Answer}$$

Given:
Distance NAUT. M. = 75 NM
Fuel flow = 85 Litres/Hour
Ground speed = 145 Knots

$$\frac{75 \text{ NM}}{145 \text{ Kts}} \cdots \frac{00.31}{60} \cdots \frac{44 \text{ Litres}}{85 \text{ Litres/hour}} = \text{Answer}$$

Therefore, ETI = 31 minutes and fuel required = 44 Litres.

50 – Fuel Consumption in A or GNM/lb   CPL

The fuel consumption in air or ground nautical miles per pound can be calculated based on speed or distance. Operators of larger aircraft find fuel calculations done in nautical mils per pound (or Litres) find this method more convenient.

Formula:

$$\frac{\text{TAS (or Ground Speed)}}{\text{Fuel Consumption}} \cdots \frac{\text{Air (or Ground) NM/Lb}}{10} = \text{Answer}$$

Given:
TAS (or Ground speed) = 125 Knots
Fuel consumption = 58 Pounds/hour

$$\frac{125 \text{ Kts}}{58 \text{ Pounds/Hour}} \cdots \frac{2.15 \text{ NM/Lb}}{10} = \text{Answer}$$

51 – Air and Ground NM/Lb from Consumption

From the known fuel consumption, true air speed and ground speed, the air or ground nautical miles per pound (or Litres) can be found.

Formula:

$$\frac{10}{\text{Fuel consumption}} \ldots \frac{\text{ANM/Lb}}{\text{KTAS}} \ldots \frac{\text{GNM/Lb}}{\text{Ground Speed}} = \text{Answer}$$

Given:
Fuel consumption = 58 Pounds/Hour
TAS = 125 Kts
Ground speed = 103 Kts

$$\frac{10}{\text{58 Lb/Hour}} \ldots \frac{\text{2.15 ANM/Lb}}{\text{125 Kts}} \ldots \frac{\text{1.77 GNM/Lb}}{\text{103 Kts}} = \text{Answer}$$

52 – GNM/Lb from Distance and Fuel Used   CPL

Another formula for calculating GNM/Lb from distance gone and fuel used. This formula is a variation on the previous two formulas.

Formula:

$$\frac{\text{Distance NM}}{\text{Fuel Used}} \dots \frac{\text{GNM/Lb}}{10} = \text{Answer}$$

Given:
Distance   = 460 NM
Fuel used   = 360 Lb

$$\frac{460 \text{ NM}}{360 \text{ Lb}} \dots \frac{1.27 \text{ GNM/Lb}}{10} = \text{Answer}$$

53 – Consumption in NM/1000 lb ATPL

For large multi-engine transport aircraft, the fuel consumption maybe measured in NM/1000 Lb of fuel (NM/1000 Lb) or NM/1000 Kg.

Place the TAS over the fuel consumption and read the answer on the outer scale over the '10 Index'. This formulas is the same as the previous one except the answer is expressed in NM/1000 Lb (or Kg) of fuel used.

Formula:

$$\frac{KTAS}{Fuel\ Consumption} \cdots \frac{NM/1000\ Lb\ (Kg)}{1000\ Lb\ (10)} = Answer$$

Given:
Consumption = 11 500 Lb/Hour
KTAS = 475 Kts

$$\frac{475\ Kts}{11500} \cdots \frac{41.3\ NM/Lb}{10} = Answer$$

54 – Conversion NM/Lb to Statute Miles/Lb   CPL

For conversions between nautical miles per pound and statute miles per pound. The Nautical Mile/Statute Miles conversion scale is located above the outer scale adjacent to the number 54 to 62 and is used for these conversions. The location may differ on other makes of E6-B's.

Method:
- Set NM/Lbs against the Naut. M index.
- Under Stat. M index, read off the Statute Miles/Lb

Example:

Reference	Naut. M	Stat. M
Answer	= 1.37 NM/Lb	1.58 SM/Lb

55 – Fuel-stop Ground Speed PPL

The fuel-stop ground speed is the minimum ground speed that must be achieved to complete the flight without using the reserve fuel. This is a convenient formula to use if head-winds are expected on the proposed flight, or it the flight time is near the maximum endurance.

Formula:

$$\frac{\text{Total Distance}}{\text{Safe endurance}} \cdots \frac{\text{Fuel-stop Ground Speed}}{60} = \text{Answer}$$

Given:
Total distance = 350 NM
Safe endurance = 03.00

$$\frac{350 \text{ NM}}{03.00} \cdots \frac{116 \text{ Kts}}{60} = \text{Answer}$$

56 – Contingency Reserve Percentage   * ATPL

Regulations may require a percentage of flight plan fuel to be added to the fuel reserves. Over Reserve Percentage find the extra amount of Reserve Fuel to be added to the Flight Plan Fuel. Over Total % read Total Fuel Required or flight plan fuel plus reserve percentage. Alternatively, given total fuel available over total percentage, read Flight Plan Fuel remaining over the '10 Index'.

Formula:

$$\frac{\text{Reserve Fuel}}{\text{Reserve \%}} \cdots \frac{\text{Flight Plan Fuel}}{\text{100\% (10 Index)}} \cdots \frac{\text{Total Fuel Required}}{\text{Total \%}} = \text{Answer}$$

Given:
Flight fuel = 475 Litres
Reserve % = 15%

$$\frac{\text{71 Ltrs}}{15\%} \cdots \frac{\text{475 Ltrs}}{100\%} \cdots \frac{\text{546 Ltrs}}{115\%} = \text{Answer}$$

Note: The slide rule shows:
Flight plan fuel   = 475 Litres
Reserve fuel   = +71 Litres
Total fuel required   = 546 Litres

57 – Fuel Percentage Adjustment

When flight plan fuel figures require a small percentage adjustment (for temperature corrections, etc) this formula gives a more accurate answer than the previous formula, No 56 – Contingency Reserve Percentage. The given percentage is converted to decimals.

Formula:

$$\frac{\text{Total Fuel (Lb or Kg)}}{10} \cdots \frac{\pm\text{Adjustment}}{\%\ \text{Adjustment}} = \text{Answer}$$

Given:
Total fuel = 21 000 Lb
% Increase or decrease = +1⅔% = 1.66

$$\frac{21\ 000\ \text{Lb}}{10} \cdots \frac{+348\ \text{Lb}}{1.66} = \text{Answer}$$

Therefore, total fuel adjustment

$$\begin{array}{r} = 21\ 000 \\ \underline{+348} \\ 21\ 348 \qquad = \text{Answer} \end{array}$$

58 – Point of No Return (Ground Speed)   CPL

The point of no return (PNR) is the furthest point an aircraft can fly to, and return to the departure point, on a given quantity of fuel. The PNR is also known as the Radius of Action.

Formula:

$$\frac{\text{G/S Out} + \text{G/S Home}}{\text{G/S Home}} \cdots \frac{\text{Endurance (Minutes)}}{\text{Time to PNR}} = \text{Answer}$$

Given:

G/S out Kts	=	150
G/S home Kts	=	+175
G/S out + G/S home Kts	=	325

Endurance = 04.15 = 255 minutes

$$\frac{325 \text{ Kts}}{175 \text{ Kts}} \cdots \frac{255 \text{ Minutes}}{02.17} = \text{Answer}$$

59 – Point of No Return (Fuel)   CPL

This formula is similar to the previous one, Formula 58 – Point of No Return (G/S), except G/S out and G/S home are substituted for G NM/Lb (or Kg) out and home. Endurance is substituted for fuel quantity in Lb (or Kg).

Formula:

$$\frac{\text{G NM/Lb Out} + \text{G NM/Lb Home}}{\text{G NM/Lb Home}} \cdots \frac{\text{Fuel Quantity Lb (or Kg)}}{\text{PNR NM}} = \text{Answer}$$

Given:

		A	B
GNM/Kg out	=	0.0655	or 2.53 Lb
GNM/Kg home	=	+0.0710	+2.39 Lb
GNM/Kg out + home	=	0.1365	4.93

Fuel quantity Kg = 36 500 or 260 Lb

A) $\dfrac{0.1365}{0.0710} \cdots \dfrac{36\,500 \text{ Kg}}{900 \text{ NM}} = \text{Answer}$

B) $\dfrac{4.92}{2.39} \cdots \dfrac{260 \text{ Lb}}{126 \text{ NM}} = \text{Answer}$

60 – PNR Changed TAS & Fuel Flow

An en route engine failure will result in a lower cruise speed and a change in the fuel flow, causing a change in the position of the original point of no return (PNR). The PNR depends on the safe endurance, which in turn (because of the changing fuel flow and ground speed) depends on the PNR. Neither the PNR nor safe endurance can be found until the other is known. Therefore, we must find how much fuel will be used to fly a known distance, i.e., 100 NM out and back at the given ground speed and fuel flow. This is found by using the relevant part of formula 49 – ETI & Fuel Required:

$$\frac{\text{Distance NM}}{\text{Ground Speed}} \cdots \frac{\text{Fuel Required}}{\text{Fuel Flow}} = \text{Answer}$$

Given:

Fuel available	510 Litres
G/S out 2 engines	190 Kts
G/S home 1 engine	171 Kts
Fuel flow out	120 Ltr/hour
Fuel flow home	82 Ltr/hour

1A: $\dfrac{100 \text{ NM}}{190 \text{ Kts}} \cdots \dfrac{63 \text{ Ltr}}{120 \text{ Ltr}} = \text{Answer}$

1B: $\dfrac{100 \text{ NM}}{171 \text{ Kts}} \cdots \dfrac{48 \text{ Ltr}}{82 \text{ Ltr}} = \text{Answer}$

1. Using the formula twice, find fuel required to fly:
   100 NM out at 190 Knots uses 63 Litres, and
   100 NM home at 171 Knots uses 48 litres.
   Therefore, total fuel used 100 NM out and back is:
   63 + 48 Litres = 111 Litres.

1. With the fuel required now known to fly 100 NM out and back (111 Litres) and the fuel available (510 Litres) we can now use the formula below No 60, PNR Changed TAS & Fuel Flow, to find the new distance to the PNR.

Formula № 60, PNR Changed TAS & Fuel Flow:

$$\frac{\text{Fuel Available Ltrs}}{\text{Fuel Out \& Back}} \cdots \frac{\text{PNR NM}}{100 \text{ NM}} = \text{Answer}$$

Given:

$$\frac{510 \text{ Ltrs}}{111 \text{ Ltrs}} \cdots \frac{460 \text{ NM}}{100 \text{ NM}} = \text{Answer}$$

Cross check:

1. Time to PNR: 460 NM out at 190 Kts = 145 minutes = *Answer*

2. Fuel flow at 120 Ltrs/hour for 145 minutes = 290 Litres = *Answer*

3. Find time PNR to home: 460 NM back at 171 Kts = 162 minutes = *Answer*

4. Fuel flow at 82 Ltrs/hour for 162 minutes = 220 Litres = *Answer*

5. Therefore, 290 Litres plus 220 Litres = 510 Litres total fuel = *Answer*

8 – Centre of Gravity

Introduction
Determining the aircraft's centre of gravity location before flight is an essential part of your pre-flight planning. Moving up to heavier aircraft used on commercial flights requires changes in the loading of the aircraft and hence, re-calculation of the CG. Changes in the centre of gravity due to an 'on' or 'off' load of passengers or cargo, or fuel used, loading for a required centre of gravity or conversions between the CG in inches or millimetres aft of the datum to % MAC, etc, can all be easily solved on the E6-B computer's slide rule. The formulas can be worked in either inches or millimetres, or a combination of the two using Kg for weight and inches for the 'arms'. For example; moment = Kg/inches.

61 – Initial Centre of Gravity

The initial centre of gravity (CG) is the most commonly used of all the CG calculation problems by pilots of light aircraft. Two examples are given below, one in Imperial Units and the other in Metric, either can be used.

Formula:

$$\frac{Moments/1000}{All\ Up\ Weight} \ldots \frac{CG\ aft\ of\ datum}{10} = Answer$$

Given:

Forward limit	= 33 inches	= 837 mm
Aft limit	= 46.5 inches	= 1180 mm
MAUW	= 4000 Lb	= 1814 Kg
AUW	= 3628 Lb	= 1645 Kg
Index units	= 161.71	= 1860

On the E6-B, set the total moments, or Index Units on the outer scale over the AUW on the inner scale and read the answer over the '10 Index' as follows:

$$\frac{161.71}{3628} \ldots \frac{44.6\ inches\ aft\ of\ datum}{10} = Answer$$

$$\frac{1860}{1645\ Kg} \ldots \frac{1131\ mm\ aft\ of\ datum}{10} = Answer$$

The initial CG is calculated at 44.6 inches or 1131 mm aft of datum, which falls within the CG Range. The AUW of 3628 Lb or 1645 Kg falls below the MAUW of 4000 Lb or 1814 Kg. Therefore, both the CG and AUW are within allowable limits. Remember, these figures and the following are examples for fictitious aircraft only.

62 – New CG – Plus or Minus Weight CPL

This formula is used to find a new CG position due to additional or reduced weight.

Any change in the loading of the aircraft after calculating the initial centre of gravity, (as in the previous formula) will require the CG to be recalculated. This change will occur with in-flight fuel consumption, or on- or off-loading of passengers or cargo at en route stops. Rather than rework the initial CG problem with new figures, it is much easier and quicker to recalculate the new CG using the weight and moments for the item being changed. For the example presented here, we will assume a passenger has been off-loaded and we require finding the new CG position. The same method applies for an increase in load. The *change* in CG position is calculated first and then added to, or subtracted from, the old CG.

Formula:

$$\frac{\text{Weight to be Removed or Added}}{\text{New AUW}} \cdots \frac{\text{Change of CG}}{\text{'Arm' of Old CG}} = \text{Answer}$$

Given:

1 Passenger off-loaded	= –170 Lb	= –77.1 Kg
New AUW (3628–170 Lb)	= 3458 Lb	= 1568 Kg
'Arm' of passenger	= 73.5 inches	= 1865 mm
Old CG position	= 44.5 inches	= 1131 mm
'Arm' distance minus old CG 73.5–44.5	= 29 inches	= 734 mm

Find *change* of CG on E6-B and subtract from old CG to find new CG position.

$$\frac{-170 \text{ Lb}}{3458 \text{ Lb}} \dots \frac{1.5 \text{ Inches}}{29 \text{ Inches}} = \text{Answer}$$

$$\frac{77.1 \text{ Kg}}{1568 \text{ mm}} \dots \frac{36 \text{ mm}}{734 \text{ mm}} = \text{Answer}$$

The New CG position
> = *43 inches aft of datum* = Answer
> or, *1095 mm aft of datum* = Answer

Note: The answer is found (in this problem) by subtracting the *change* of CG from the old CG.

Old CG of 44.5 inches minus 1.5 inches (found from the formula) = new CG.

In Metric it is 1131 – 36 mm = new CG 1095 mm = Answer

Note: If weight is reduced, the CG moves forward and vice versa.

63 – Loading for a Required CG CPL

Use this formula when AUW and CG position are known.

This problem uses the same formula as Formula 62, New CG – Plus or Minus Weight. But this time the known variables given are different and so the answer is found on the inner scale of the slide rule. The problem requires finding how far aft of the datum can a given load be placed to produce a given CG location. The new CG of 43" or 1095 mm in the previous example becomes the present CG.

Formula:

$$\frac{\text{Added Weight}}{\text{New AUW}} \cdots \frac{\text{CG Difference}}{\text{Distance weight} - \text{old CG}} = \text{Answer}$$

Given:

Present old CG loaded	= 43 inches	1095 mm
Required CG	= 45 inches	1143 mm
CG difference	= 2 inches	50 mm

$$\frac{+175 \text{ Lb}}{3633 \text{ Lb}} \cdots \frac{2 \text{ inches}}{41.5 \text{ inches}} = \text{Answer}$$

1. Add 'distance weight – old CG' from slide rule to present old CG to find how far aft of datum a given load can be placed.

2. Add answer to present old CG to find how far aft of datum a given load can be placed.

3. This: 41.5 inches + 43 inches = 84.5 inches aft of datum, or 974 + 1095 mm = 2069 mm aft of datum = Answer.

4. Placing 175 Lb at station 84.5" aft of datum moves the CG to the required location of 45"aft of datum

64 – Re-loading for a Required CG (Added Weight)

Loading the aircraft to produce a given or required CG location may arise when the CG is out of limits, too far aft or forward. The question is, how much additional weight can be placed at a given station to position the CG where required?

Formula:

$$\frac{\text{CG Difference}}{\text{Distance Aft of Datum}} \cdots \frac{\text{Required load}}{\text{AUW}} = \text{Answer}$$

Given:

Present CG	= 48.25 inches	= 1224 mm
Required CG	= −46.00 inches	= −1167 mm
CG Difference	= 2.25 inches	= 57 mm
Baggage Arm	= 138.7 inches	= 3520 mm
Distance Aft of Datum	= 92.7 inches	= 2352 mm
AUW	= 3510 Lb	= 1592 Kg

$$\frac{2.25 \text{ Inches}}{92.7''} \cdots \frac{85 \text{ Lb}}{3510 \text{ Lb}} = \text{Answer}$$

$$\frac{57 \text{ mm}}{2352 \text{ mm}} \cdots \frac{39 \text{ Kg}}{1592 \text{ mm}} = \text{Answer}$$

1. Placing 85 Lb (39 Kg) at station 92.7 inches (2352 mm) will move the CG forward to the required location of 46 inches (1167mm) aft of datum.

2. New AUW = 3510 Lb + 85 Lb =3595 LB = *Answer*
   Or, 1592 Kg + 38 Kg = 1630 Kg = *Answer*

65 – CG Change Due to Weight Shift

Use this formula to find the change of CG location when a given weight is shifted a given distance. The AUW remains the same because weight is not added or removed but shifted within the aircraft.

Formula:

$$\frac{\text{Weight shifted}}{\text{AUW}} \cdots \frac{\text{CG change}}{\text{Distance Weight Shifted}} = \text{Answer}$$

Given:

Weight Shifted	= 65 Lb	= 29.5 Kg
AUW	= 3250 Lb	= 1474 Kg
Distance Weight to be Shifted	= 90 inches	= 2284 mm

$$\frac{65 \text{ Lb}}{3250 \text{ Lb}} \cdots \frac{1.8 \text{ Inches}}{90 \text{ Inches}} = \text{Answer}$$

$$\frac{29.48 \text{ Kg}}{1474 \text{ Kg}} \cdots \frac{45 \text{ mm}}{2284 \text{ mm}} = \text{Answer}$$

Moving weight forward, subtract CG Change (E6-B answer) from the old CG.

Moving weight aft, add CG change (E6-B answer) to old CG.

66 – Weight to be Shifted CPL

This formula is used to find the weight of a given load to be shifted when the CG difference (distance from present to required new CG, and the distance the weight to be shifted and AUW are given, The formula here is the same as Formula 65, CG Change Due to Weight Shift, but with each side of the equation transposed.

Formula:

$$\frac{\text{CG Difference}}{\text{Distance Weight Shifted}} \ldots \frac{\text{Weight Shifted}}{\text{AUW}} = \text{Answer}$$

Given:

CG Difference	= 1.35 inches	= 34.25 mm
Distance Weight Shifted	= 105 inches	= 2664 mm
AUW	= 3345 Lb	= 1517 Kg

$$\frac{1.35 \text{ Inches}}{105 \text{ Inches}} \ldots \frac{43 \text{ Lb}}{3345 \text{ Lb}} = \text{Answer}$$

$$\frac{34.26 \text{ mm}}{2664 \text{ mm}} \ldots \frac{19.5 \text{ Kg}}{1517 \text{ Kg}} = \text{Answer}$$

67 – CG Aft of LEMAC from % MAC ATPL

For larger aircraft, the centre of gravity is usually expressed as a percentage of the mean aerodynamic chord (% MAC) although it also can be applied to all aircraft of any size.

Most pilots use the inches or millimetres aft of datum method. However, if for any reason you need to change from inches or millimetres to % MAC, this can be done on the E6-B using either of the two following formulas. Use the formula below to find the CG in inches or millimetres aft of datum when given the CG as % MAC and LEMAC stations.

Note: LEMAC is the abbreviation for leading edge (of the) mean aerodynamic chord.

Formula:

$$\frac{MAC}{10} \cdots \frac{CG \text{ Aft of LEMAC}}{CG \text{ \% MAC}} = Answer$$

Given:

MAC	= 152 inches	= 386 mm
LEMAC	= 245 inches	= 6218 mm
CG Location	= 18%	= 18%

$$\frac{152 \text{ Inches}}{10} \cdots \frac{27.4 \text{ Inches}}{18\% \text{ MAC}} = Answer$$

$$\frac{386 \text{ mm}}{10} \cdots \frac{695 \text{ mm}}{18\% \text{ MAC}} = Answer$$

Note: Add E6-B answer 27.4 inches (695 mm) to LEMAC 245 inches (6218 mm) to find the CG Aft of LEMAC:

27.4 inches + 245 inches
= 272.4 inches aft of LEMAC = *Answer*

695 mm + 6218 mm = 6913 mm aft of LEMAC = *Answer*

68 – CG % MAC from LEMAC

In this formula we are looking to find the CG in % MAC given the MAC and the CG location aft of LEMAC. This formula is set out the same as Formula 67, CG Aft of LEMAC from % MAC, but this time the answer is found on the inner scale of the E6-B slide rule.

Formula:

$$\frac{MAC}{10\ Index} \cdots \frac{CG\ aft\ of\ LEMAC}{CG\ \%\ MAC} = Answer$$

Given:

MAC	= 155 inches	= 3934 mm
100%	= 10 Index	= 10 Index
CG Aft of LEMAC	= 24.3 inches	= 616 mm

$$\frac{155\ Inches}{10} \cdots \frac{24.3\ Inches}{15.7\%} = Answer$$

$$\frac{3934\ mm}{10} \cdots \frac{616\ mm}{15.7\%} = Answer$$

Note: Answer is on the inner scale of the E6-B.

69 – Converting CG % MAC Limits

The aircraft's CG limits given in % MAC can easily be converted to inches or millimetres and vice versa using this formula. This is the same formulas the previous one CG in % MAC from LEMAC. Both the fore and aft limits are read off the E6-B with only one setting.

Formula:

$$\frac{MAC}{10\ Index\ (100\%)} \cdots \frac{Forward\ Limit}{Forward\ Limit\ \%\ MAC} \cdots \frac{Aft\ Limit}{Aft\ Limit\ \%\ MAC} = Answer$$

Given:

MAC	= 152 inches	= 3857 mm
100%	= 10 Index	= 10 Index
Fwd Limit % MAC	= 11%	= 11%
Aft Limit % MAC	= 25%	= 25%
LEMAC	= 245 inches	= 6218 mm

$$\frac{152\ Inches}{10} \cdots \frac{16.7\ Inches}{11\%} \cdots \frac{38\ Inches}{25\%} = Answer$$

$$\frac{3857\ mm}{10} \cdots \frac{423\ mm}{11\%} \cdots \frac{964\ mm}{25\%} = Answer$$

1. Add answers to LEMAC 245 inches to find the limits in inches or millimetres aft of datum (MAC x given % in decimal form).

2. The forward limit becomes 245 + 16.7
   = 261.7 inches = Answer 1
   The aft limit is 245 + 38 = 283 inches = Answer 2

Note: If converting fore and aft limits from inches (or mm) to % MAC, the answer will be on the inner scale.

70 – Empty Weight CG

Finding the empty weight centre of gravity is a problem pilots will not normally be called upon to calculate; this is usually done by the aircraft manufacturer or ground maintenance engineers. But this formula is included to complete this series on CG calculations.

The aircraft must first be weighed to find the weight on the nose-wheel (or tail-wheel) and on the main-wheels to find the empty AUW. The E6-B formula finds the distance of the CG from the main-wheel centreline. To find this answer we must add the given distance from the datum to the main-wheel centreline in order to determine the aircraft's empty weight CG. For nose-wheel aircraft, the E6-B answer is subtracted from the 'datum to main-wheel distance' and for tail-wheel aircraft, the answer is added.

Formula:

$$\frac{\text{Tail-wheel Weight}}{\text{Empty Weight}} \cdots \frac{\text{Distance} - \text{Main-wheels to CG}}{\text{Distance} - \text{Main-wheels to Tail-wheel}} = \text{Answer}$$

Given:

Nose-wheel weight	= Nil	= Nil
Tail-wheel weight	= 65 Lb	= 29.4 Kg
Aircraft empty weight	= 1252 Lb	= 568 Kg
Distance datum to main-wheels	= 4.5 in.	= 114 mm
Distance mains to tail-wheel	= 217 in.	= 5507 mm
Weight on main-wheels	= 1187 lb	= 538.4 Kg

$$\frac{65 \text{ Lb}}{1252 \text{ Lb}} \cdots \frac{11.25}{217 \text{ Inches}} = \text{Answer}$$

9 – Wind Slide Computer

Introduction to the Wind Slide Computer

The first eight sections of this book dealt with the circular slide rule side of the E6-B computer. This section now deals with the Wind Slide Computer on the reverse side. It is used mainly for solving problems associated with the triangle of velocities – true heading, TAS, track, ground speed and wind velocity, etc. Given any two sets of figures, the third can be found.

Different methods for finding the wind velocity (W/V) while in flight follows, and also solving the average W/V for use in flight-planning the climb. Head-wind and tail-wind components follow, along with the Air Plot Diversion, DR Error and many more. Finally ending with what must be two of the most complicated formulas in this book, The Point of Safe Diversion and Ship Interception. Follow the methods through, they are quite simple really!

Solving problems on the Wind Slide Computer can be divided into five basic methods:

- Centre Line as the Air Vector (TH & TAS)
- Centre Line as Ground Air Vector (TR & G/S)
- Wind Velocity on Climb
- Wind Slide Navigation
- Square Grid problems.

Centre Line as Ground Vector (GV)

When the centre line is used as the ground vector, the wind-cross is drawn above the grommet. Therefore, the wind direction is from the TH & TAS vector on the drift line and speed arc, towards the TR & G/S on the centre line. The drift is read directly from the position of the wind cross on the drift line, without any need for further adjustment. This method is preferable in the flight-planning stage and for calculating new headings during flight. If the wind cross is on the left of the centre line, drift is to the right and vice versa. Therefore, the aircraft heading is turned towards the wind cross to correct for drift. This is the method preferred by professional pilots.

In all wind-finding methods in this book, the wind-cross is always placed below the grommet to read off the wind velocity found. It follows, having found the wind velocity the cross will need to be transferred to above the grommet to continue using the centre line as the ground vector. Only a small adjustment to the slide and rose is then required to find the next TH & G/S.

Centre Line as Air Vector (AV)
At times during flight, it maybe more convenient to use the wind slide computer with the centre line as the air vector (TAS & TH). With this set-up, the wind vector acts from the TH &TAS on the centre line to the TR & G/S on the drift line and speed arc. With the wind cross placed below the grommet, the slide and compass rose will need to be adjusted to match the drift indicated on the slide with that on the rose; this is more time consuming than the ground vector method.

When using both methods of ground vector and air vector, be aware you do not confuse which method you are using to solve a given problem. Drawing an arrow head in the wind direction is a good reminder.

Note:

- When each formula is presented in this book, it will be tagged on the right-hand side of the page in line with the title stating if it is a ground vector or, air vector problem thus:

    Air vectors are tagged as '**AV**'.
    Ground vectors are tagged as '**GV**'.

Note: Contrary to a pilot's normal thinking of the ground below and air above, the air vector and ground vector are reversed thus:

- Ground vector = wind cross above.
- Air vector = wind cross below.

Square Grid Problems

The square grid (or graph) is located at the bottom of the Low-speed wind slide. It has several uses, which are all covered at the end of the wind problem section.

- The wind vector problems that follow do not have formulas as such; rather they are presented as 'methods' on how to set up the wind slide computer.

Centre Line as Ground Vector

71 – True Heading & Ground Speed (1)   GV/PPL

This is one of the most common vector triangle problems, requiring the calculation of the true heading (TH) and ground speed G/S during pre-flight planning.

Given:
W/V = 210/28
TAS = 130 Kts
Track = 175° True

Method:

1. Set wind direction under TH index 210°

2. Mark wind cross *above* grommet at 28 Knots

3. Rotate compass rose to set TR 175° under TH index

4. Place wind cross over speed arc at TAS 130 Knots

5. Read off drift at 7° left

6. Add 7 drift to TR 175 Knots to find TH 182° = *Answer 1*

7. Set 182° under TH index and read off G/S under grommet 105 Knots = *Answer 2*

72 – True Heading & TAS (1)

Given the W/V, required TR and G/S, the true heading and TAS can be found using the same method as the previous two problems. The wind cross is placed above the grommet.

Given:
W/V = 120/25
Track = 170° True
G/S = 175 Kts

Method:

1. Set W/V 120/25 above the grommet

2. Set the grommet over the speed arc for G/S 175 Kts

3. Set TR 170° under the TH index

4. Read off wind correction angle (WCA) 5° Port (left)

5. Read off TH (170°–5°) = 165° under wind cross = *Answer 1*

6. Read off TAS 194 Kts under wind cross = *Answer 2*

73 – Radius of Action – TH & G/S   GV/CPL

This one is the same method as that used for finding the TH and G/S. The only difference here is the TH and G/S are found for outbound and return tracks. The G/S is then entered into the formula for determining the Radius of Action.

Note: Make a pencil mark at the 180° position on the fixed frame around the compass rose.

Given:
Outbound track = 292°
Wind velocity = 330/25
TAS = 210 Knots

Method:

1. Set W/V 330/25 *above* grommet

2. Set outbound track 292° under TH index

3. Place wind cross over TAS 210 Kts

4. Read off G/S out under the grommet 190 Kts = Answer 1

5. Read off drift at 4° left to find outbound TH 296° = Answer 2

6. Set reciprocal track of 112° under TH Index

7. Repeat steps 3 to 5 for return track

8. Read return G/S 228 Kts = Answer 3

9. Read return TH 108° = Answer 4

74 – Simple Diversion GV/CPL

The track and distance direct to an alternate airfield from the present position can be found by using simple trigonometry on the wind slide.

In this example, the alternate airfield is further on past the destination airfield. But if the alternate airfield is closer, a point will be reached where both the destination and alternate are the same distance to fly to. This will be indicated on the wind vector slide when the grommet and wind cross are both on the same speed/distance arc. Having found the TR and distance to the alternate airfield, TH and G/S must then be found.

Given:
Distance, destination to alternate = 50 NM
Track, destination to alternate = 072°
Distance to destination = 110 NM
Track to destination = 020°

Method:

1. Turn compass rose to place Track from destination to alternate under TH index = 072°

2. Mark pencil cross at 50 NM above grommet

3. Turn compass rose to place track to destination under the TH index= 020°

4. Set the grommet over the 'distance to go' to destination = 110 NM

5. Read distance from present position to alternate airfield = 146 NM = Answer 1

6. Read track direct to alternate airfield = 020° +016° = 036° = Answer 2

Centre Line as Air Vector

75 – True Heading & Ground Speed (2)   AV/PPL

This is an alternative method of finding TH and G/S using the centre line as the Air Vector. The disadvantage with the wind cross being placed *below* the grommet is the need to juggle the compass rose to match the wind drift on the plotting disc with the difference between the TH and track on the Wind Drift Correction Scale. But, the advantage here is having found the W/V from the following methods, the TH and G/S can be found with a simple adjustment to the slide and compass rose.

Given:
W/V = 210/28
TAS = 130 Kts
Track = 175° True

Method:

1. Set W/V 210/28 *below* the grommet

2. Set required track 175° under the TH index

3. Place grommet over TAS 130 Kts

4. Read off drift 9° left

5. Juggle rose to set drift under TH index same as on drift line 7° left. (TR 175° + 7° = 182°)

6. Read off TH 182° under TH index = Answer 1

7. Read off G/S 105 Kts under wind cross = Answer 2

76 – True Heading & TAS (2)

When the mission calls for an arrival at a given location and at a given time, the required track, G/S and W/V will be known. The problem remains to find at what TH and TAS to fly. Remember, having found the required TAS, this must be converted back to IAS.

Given:
W/V = 090/20
Track = 240°
G/S =140 Kts

Method:

1. Set W/V 090/20 *below* the grommet

2. Set TR 240° under TH index

3. Adjust slide to place wind cross on speed arc, G/S 140 Kts

4. Read approximate drift 4° right, at end of wind cross

5. Adjust slide to correct drift 5° right

6. Read TH 235° under TH index = Answer 1

7. Read TAS 123 Kts under grommet = Answer 2

77 – Track & Ground Speed AV/PPL

The track made good and ground speed can be found after flying a constant heading with the TAS and W/V known, where no fix is available. The wind cross is drawn *below* the grommet.

Given:
W/V = 175/15
TAS = 145 Kts
TH = 230°

Method:

1. Set W/V 175/15 *below* the grommet

2. Rotate rose to set TH 230° under the TH index

3. Place grommet on TAS 145 Kts

4. Under wind cross read G/S 136 Kts = Answer 1

5. Under wind cross read drift 6° right; therefore, TR is 230° + 6° = 236° = Answer 2

78 – Wind Velocity from Drift & Ground Speed

During flight, the values of drift and ground speed are frequently updated. These values, combined with known TAS and TH can be used to find the W/V affecting the aircraft over the time of observation.

Given:
Track = 340°
G/S = 204 Knots
TAS = 225 Knots
Drift = 7° right

Method:

1. Set TR 340° under TH index

2. Set 225 Kts under grommet

3. Mark pencil line for 7° right drift along 7° drift line to right of centre line

4. Mark G/S 204 Kts on speed arc across 7° right drift line

5. Read off wind direction 293° under TH index = Answer 1

6. Read off wind speed 35 Kts (difference between grommet and wind cross) = Answer 2
   Therefore W/V is 293/35.

79 – Wind Velocity from Triple Drift CPL

The wind velocity can be found after flying on three different headings with known drift readings and constant true air speed. For accuracy, the headings must be more than 30° apart, preferably about 60° difference.

Given:
TAS = 165 Kts
1st. heading = 150°, drift 10° right
2nd. heading = 195°, drift 6° right
3rd. heading = 110°, drift 8° right

Method:

1. Set grommet over TAS 165 Kts

2. Rotate rose to TH 150° under TH index

3. Mark pencil line parallel to drift line 10° right

4. Repeat steps No 2 and 3 for the other two headings

5. Rotate rose to place intersecting drift lines (wind cross) *below* grommet

6. Read off W/V = 060/28 = Answer

80 – Wind Velocity from Triple Ground Speed PPL

This formula is basically the same as for 79 – Triple Drift, except ground speeds are used instead of drift.

Given:
TAS = 235 Kts
1st. heading = 120°, G/S 250 Kts
2nd. heading = 165°, G/S 215 Kts
3rd. heading = 085°, G/S 274 Kts

Method:

1. Set 235 Kts below grommet

2. Set TH 120° under TH index

3. Draw arc on 250 Kts speed arc

4. Repeat for 2nd. and 3rd. headings

5. Rotate compass rose to place wind cross *below* the grommet (intersection of arcs)

6. Read off W/V = 225/48 = Answer

81 – Wind Velocity from Drift & Wind Lanes   CPL

When flying over water, the wind direction can be found from the wind lanes visible on the water when flying below 2500 feet ASL. The wind direction veers with altitude so add 10° veer to the direction of the wind lanes.

Given:
Wind lanes bear = 320°T – 140°T
Veer +10° = 330°T – 150°T
TAS = 145 Kts
TH = 255°
Drift = 9° left

Method:

1. Rotate rose to wind direction 330° True under TH index

2. Mark pencil line above and below grommet along centre line

3. Rotate rose to aircraft's TH 255°

4. Mark pencil line along 9° left drift line

5. Rotate rose to place wind cross *below* grommet

6. Read off W/V = 330/24 = Answer

Wind Velocity on Climb

82 – Arithmetical Mean Wind Velocity CPL

During flight planning, an average wind velocity is required for computations on the climb or descent legs of the planned trip. Three methods are presented here. The first method known as the Arithmetical Mean Wind Velocity, assumes only a small change in wind direction and velocity with an increase of altitude, and a constant rate of climb, which is limited to the lower altitudes. The average W/V is found by totalling all the given wind directions and speeds, and dividing each by the total number of terms.

Given:
1000 feet W/V = 270/10
3000 feet W/V = 280/15
5000 feet W/V = 295/20
7000 feet W/V = 310/20
Total units = 1155/65

1155 ÷ 4 = 288° and 65 ÷ 4 = 16

Therefore, the average W/V = 288/16 = Answer

83 – Vector Average Wind Velocity

This second method of averaging the W/V for climb produces better results and is more accurate than the previous formula. It is used when the W/V has a greater change in direction and speed. As in the first method, its accuracy is limited to the lower altitudes and a constant rate of climb. Another advantage is; this method can also be used to find the average W/V over the planned route.

Given:
1000 feet W/V = 250/10
3000 feet W/V = 280/15
5000 feet W/V = 320/20
7000 feet W/V = 340/25

Method:

1. Set 250° under TH index, and place grommet on top line of the square grid

2. Draw line vertically *down* from grommet a suitable number of squares to equal 10 Knots

3. Rotate rose to 280° under TH index

4. Again, draw line vertically down from end of first vector, parallel to the centre line equal to 15 Knots

5. Repeat above steps for each subsequent altitude and given W/V

6. Rotate rose to place end of last vector on centre line *below* grommet

7. Read number of units and divide by number of vectors, i.e. 70 Kts ÷4 = 18 Kts wind speed = Answer 1

8. Under TH index, read 'Vector Mean' wind direction =308° = Answer 2
   Therefore, Vector Average W/V = 308/18 = Answer 1 and 2

Note: See also Formula 25, 'Weighted Wind Components'.

84 – Mean Equivalent Wind Velocity CPL

The Mean Equivalent Wind Velocity method is better suited when the W/V is irregular and the rate of climb varies with altitude. Because of the reducing rate of climb, the time spent in each height band has to be taken into consideration. This is done by multiplying the wind-speed by the time spent in the height band. The square grid is then used to plot the resulting vector units for each height band. Each vector is plotted vertically down *from* the end of the previous vector, as in the Vector Average method. The total vector units are then divided by the total time spent in the climb, to give the wind speed.

Given:

W/V	Height	Height Band	Time in Band	Vector Units
160/15	3000 Ft	1000 to 4000 Ft	2 minutes	15×2=30
175/25	5000 Ft	4000 to 7500 Ft	2½ minutes	25×2.5=62
190/40	9000 Ft	7500 to 12500 Ft	4 minutes	40×4=160
220/50	15000 Ft	12500 to 17000 Ft	4½ minutes	50×4.5=225
245/65	19000 Ft	17000 to 21000 Ft	4 minutes	65×4=260

Method:
1. Set 160° (3000 Ft wind direction) under THI index and place grommet on the top line of the square grid
2. Mark a cross 30 vector units below grommet to a suitable scale, i.e. 1 small square = 10 units
3. Rotate rose to next wind direction of 175°

continued overleaf

1. Adjust slide to place first cross on top line of square grid
2. Draw in next cross for 62 units vertically below first cross
3. Repeat above steps for each wind direction and units
4. Rotate rose to place last wind cross on centre line.
5. Adjust slide to place grommet on the top line
6. Under TH index, read Mean Equivalent wind direction = 210° = Answer 1
7. Determine total number of units from the grommet to the last wind cross = 740 units
8. Divide number of units by the time spent on the climb, 740 ÷ 17 = 43 Kts = Answer 2
9. Therefore, the Mean Equivalent W/V = 210/43 = Answer 3

Wind Slide Navigation

85 – Off-track Correction

When a fix determines the number of miles an aircraft is off-track, the track error and correction angle can be determined using the square grid on the wind slide. This is an alternative method to that shown by the '1 In 60' Formula 40, Off-track Correction.

Given:
Total distance = 145 NM
True heading = 035° True
Distance off track = 5 NM left
Distance gone = 52 NM

Method:
1.   Use the square grid. One square = 1NM
2.   Set TH 035° under TH index
3.   From grommet, draw line 5 units left of centre line (5 NM left of track)
4.   Adjust slide to place grommet on distance gone, 52 NM
5.   Find angle off track 6° left, on drift line; this is track error
6.   Adjust slide on distance to go, 93 NM
7.   Read angle to intercept rack 3° left on drift line, correction angle
8.   Add track error and correction angle (6° + 3° = 9° to the right)
9.   Correction to initial heading is 035° + 9° = 044° True = *Answer*

Note: See also Formula 40, Off-track Correction

86 – Point of Safe Diversion

This formula is also known as the 'Point of No Alternate' and also 'Radius of Action to an Alternate Base'.

The Point of Safe Diversion (PSD) is the farthest point along a route where the aircraft can divert to an off-track alternate airfield while maintaining the required fuel reserves. The same formula and method is used where the aircraft is required to fly outbound as far as possible and return to a different base. The problem requires solving the relationship between 'rates of departure' (S1) 'rates of closure' (S2) and 'safe endurance', along the line of constant bearing between the departure point and the alternate base.

The wind slide computer is used for plotting the first part of the problem and the circular slide rule is used to solve the formula. However, take note of Formula 86, Point of Safe Diversion, used below, it is similar to the radius of action formula, except rate of departure (S1) and rate of closure (S2) are used in place of ground speed out and home; Same method, different terminology. Use ground speed out (G/S 1) to find time to turn.

Given:

True track out	= 295°
W/V	= 240/30
True air speed	= 180 Kts
Fuel endurance	= 03.30 = 210 minutes
Alternate base:	
Bearing from departure	= 025°
Distance from departure	= 172 NM

Find:
TH out
G/S out (GS 1)
Rate of departure (S1)
TH to alternate base
G/S to alternate base
Rate of closure (S2)
Time to turn (T)
Radius of action

Method:

1. First, calculate the hourly increment by dividing the distance (172 NM) from the departure point to the alternate base by fuel endurance in hours (3.5) = 172/3.5 = 49 Knots

2. Use square grid, set W/V 240/30 *below* the grommet, and label as 'W'.

3. Set bearing to alternate 025° true, under the TH index

4. Adjust slide to place the end of the wind vector (W) on the top line of square grid

5. Draw line vertically down from the end of the wind vector with length equal to the hourly increment 49 Knots, label as 'R' for relative wind vector

6. Set track out 295° under TH index

7. Draw line vertically down from the end of the wind vector (W)

8. Adjust slide to place TAS 180 Knots under the grommet

9. Rotate compass rose to place the line drawn vertically from 'W' parallel to the nearest drift line

10. Read TH 287° under TH index = *Answer 1*

continued overleaf

1. Read G/S out 162 Knots at head of vector = *Answer 2*

2. Read rate of departure (S1) at end of hourly increment (R) 170 Knots = *Answer 3*

3. Draw line parallel to drift line from point (R)

4. Rotate compass rose approximately 180° to place line just drawn from point (R), parallel to drift line on opposite side of centre line

5. Read TH 091° to alternate under TH index = *Answer 4*

6. Read G/S 207 Knots to alternate airfield at end of wind vector (W) = *Answer 5*

7. Read rate of closure (S2) 187 Knots at end of hourly increment (R) = *Answer 6*

8. Use S1 170 Knots and S2 187 Knots plus endurance 210 minutes in the formula below to find the time to turn.

Note: This is the same formula as Formula 19 – Radius of Action, renamed here as:

Formula 86, Point of Safe Diversion

$$\frac{S1 + S2}{S2} \cdots \frac{Endurance\ (Minutes)}{Time\ to\ Turn} = Answer$$

$$\frac{357}{187} \cdots \frac{210}{01.50} = Answer\ 7$$

Calculate the radius of action using Formula 9 – Time, Speed & Distance

Given:
Time   = 01.50
G/S 1 (out)   = 162 Knots

$$\frac{162\ Knots}{60} \cdots \frac{210}{01.50} = Answer\ 8$$

The Ship Interception formula is used mainly by Naval pilots carrying out shore to ship carrier on board delivery (COD's) operations, or any operation calling for finding a ship out at sea.

When intercepting a moving base, the ship's track (course for a ship) and its speed must be taken into account as the fourth vector of relative motion. Also required is the ship's bearing and distance from the departure airfield. This formula is less involved than the Point of Safe Diversion, but the method is similar. We only need to find the required details for the outbound trip to the ship (no return).

Given:

W/V	= 225/25
Ship's track	= 210°
Ship's speed	= 32 Knots
Ship's bearing	= 133°
Ship's distance	= 95 NM
Aircraft TAS	= 180 Knots

Find:
TH to intercept
Track to intercept
G/S to intercept
Time to intercept
Rate of closure (S1)

Method:

1. Using the square grid, set W/V 225/25 below the grommet and label 'W'

2. Set ship's track 210° under TH index

3. Adjust slide to place top line of square rid on end of wind vector 'W'

continued overleaf

1. Draw line vertically down from end of W/V 'W' with length equal to ship's speed 32 Knots, label 'R' for relative wind vector

2. Set ship's bearing 133° under TH index

3. Draw line vertically down from end of ship's speed 'R'

4. Adjust slide to place aircraft's TAS 180 Knots under grommet

5. Rotate compass rose to place the line drawn vertically from 'R' parallel to nearest drift line

6. Under TH index, read TH 153° to intercept ship = *Answer 1*

7. Read G/S 174 Knots to intercept at the end of the wind vector 'W' = *Answer 2*

8. Read drift 8° left at end of wind vector. Therefore, to intercept ship, the aircraft's TR 145° = *Answer 3*

9. Read rate of closure (S1) 164 Knots at end of ship's vector 'R' = *Answer 4*

10. With values found, enter into formula below to find time to intercept. This formula is similar to Formula 9, Time, Speed & Distance.

Formula:

$$\frac{(S1)\ \text{Rate of Closure (Kts)}}{60} \cdots \frac{\text{Distance to Close (NM)}}{\text{Time to Intercept}} = \text{Answer 5}$$

$$\frac{(S1)\ 164\ \text{Kts}}{60} \cdots \frac{95\ \text{NM}}{35\ \text{minutes}} = \text{Answer 5}$$

88 – Convergency (Graphical Method)   ATPL

Convergency (on Lambert's chart) or Conversion angle (on Mercator's chart) needs to be applied when plotting radio bearings on these charts at distances over 600 NM. Calculating convergency or conversion angle without using trigonometry tables can be done graphically on the wind slide. Remember, the difference of longitude is halved when calculating the conversion angle on Mercator charts. See also Formulas No 18 & 19.

Given:
Latitude   = 35°
Difference of longitude = 21°

Method:

1. Set compass rose to 360° under TH Index

2. Draw a line from the grommet up the centre line

3. Turn compass rose to given latitude 35°

4. Adjust slide to place grommet over the difference of longitude 21° on square grid, counting down from top line

5. Where pencil line intersects top line of grid, read off Convergency 12° = Answer

89 – DME Off-track Corrected Reading

A simple and quick method to compute the distance on track to a point abeam an off-track DME beacon follows:

Given:
DME read-out   = 65 NM
Track difference   = 25°

Method:

1. Set the grommet over the speed arc for the DME distance of 65 NM

2. Align a straight edge (pencil) across the disc with the distance/speed arc and drift line at 25° on left and right-hand sides of the centre line

3. On centre line read off the distance on track to abeam DME beacon 59 NM = Answer

Note: These are the same figures as used in Formula 16, Off-track DME Distance.

Square Grid Problems

90– Wind Velocity from Airborne Radar ATPL

A method for finding wind velocity when airborne can be achieved by using airborne radar or taking two bearings off the same beacon with a radio nav aid (VOR & DME). The two bearings and distances to the side of the aircraft's track and the time interval between bearings are used to find the distance run, track made good and ground speed between the bearings.

Given:
Time interval = 17 minutes
1st. bearing = 077°, 104 NM, time 14.32
2nd. bearing = 118°, 86 NM, time 14.49
True air speed = 214 Kts
True heading = 032°

Method:

1. Set grommet on top line of square grid

2. Rotate rose to first bearing of 077°

3. Using a suitable scale (i.e. 1 small square = 2 NM) mark off distance 104 NM below grommet

4. Rotate rose to second bearing of 118°

5. Mark off distance 86 NM *below* grommet

6. Rotate rose to place both bearing crosses on the same vertical line

7. Under TH, read track made good (TMG) 022° between the two bearings

8. Find distance 68 NM between the bearing crosses, (34 × 2 = 68 NM). Note: The W/V can be found by using Formula 78, Wind Velocity from Drift & Ground Speed, using the data found from above. *continued overleaf*

115

Data found:

Track made good   = 022°
Distance gone   = 68 NM in 17 minutes
Ground speed   = 240 Knots
Drift   = 10° left
True heading   = 032°
TAS   = 214 Kts,

Therefore, W/V 150/48 = Answer

91 – Correcting Wind Velocity by DR Error   ATPL

When the observed fix is different to the DR position, either a wrong W/V was used in the calculation, or the W/V has changed. The correct W/V can be found by adding a correction vector, expressed as a velocity (between the DR position and the fix) to the end of the wrong W/V used. The correct W/V will then run from the origin of the wrong W/V to the fix position.

Given:
Used W/V = 270/25
DR error = 7 NM at 350°
Elapsed time = 15 minutes

Method:

1. Set wind velocity used 270/25 below grommet on square grid.

2. Rotate rose to place 350° on TH index

3. Calculate DR error per hour, 7 NM in 15 minutes = 7 x 4 = 28 NM

4. Draw pencil cross at 28 NM *up* from end of W/V vector

5. Place wind cross 28 units *below* top line of square grid

6. Rotate rose to place end of correction vector on centre line below the grommet

7. Read off correct W/V = 218/34 = Answer.

92 – Head & Cross-Wind Components (Method A)
CPL

The square grid is used here to find the head- and cross-wind components acting on the runway, when the W/V and runway direction are known. The head-wind component is used for calculating the aircraft's take-off and landing distances, while the cross-wind component on the runway must not exceed the demonstrated cross-wind component for the aircraft type.

Given:
Runway heading = 290°
Wind velocity = 260/25

Method:

1. Set the wind direction under the TH index 260°

2. Place grommet on the top line of the square grid

3. Mark cross *below* grommet for wind speed = 25 Kts

4. Rotate rose to runway heading 290°

5. Under wind cross read head-wind component = 22 Kts = Answer 1

6. Under wind cross read cross-wind component = 13 Kts = Answer 2

Note: The head-wind component is read vertically down the square grid and the head-wind component is read horizontally across the grid.

93 – Head & Cross-wind Component (Method B)

This formula is an alternative method for calculating head & cross-wind Components on an E6-B without a square grid. Although the method is not as direct as the previous square grid method, the same results are still attainable.

Given:
Runway heading = 325°
Wind velocity = 295/20

Method:

1. Draw W/V *above* the grommet 295/20

2. Rotate compass rose to runway heading 325°

3. Read off head-wind component as 17 Kts = Answer A

4. Draw horizontal line from wind-cross to centre line

5. Rotate rose to place line parallel to drift line

6. Read off cross-wind component as 10 Kts = Answer B

94 – Air Plot diversion

At times it may become necessary to divert en route on to a series of heading changes due to adverse weather or Air Traffic Control requirements, etc. The air position at the end of the diversion can be found given the TAS, W/V and the times on each heading. If greater distances or higher TAS are involved, it may be more convenient to plot the air vectors in terms of minutes rather than miles. Select a suitable scale on the square grid, e.g.: 1 square = 1 NM.

Given:
1st. heading = 270° for 5 minutes = 15 NM
2nd. heading = 300° for 6 minutes = 18 NM
3rd. heading = 320° for 4 minutes = 12 NM
Total time = 15 minutes
TAS = 180 Kts
W/V = 250/30

Method:

1. On the square grid, place the grommet on the top line

2. Set compass rose to first heading 270°, and mark line *down* 15 units (NM)

3. Set rose to second heading 300°, mark line down 18 units (NM) from end of first vector

4. Repeat operation for each change of heading

5. Adjust slide to place grommet on the top line of the square grid (start of air plot)

6. Adjust rose to place end of last air vector on the centre line, to find the bearing and distance of the zero wind position

7. Read bearing 295° under the TH index and the air distance 42 NM = Answer 1

1. When W/V is known, set wind direction under TH index 250° and mark line *above* for 7.5 units (Kts) from the end of the last heading vector

2. Turn rose to place end of wind vector on centre line

3. Under TH index, read bearing from start of air plot to end of air plot position 303° True = Answer 2

4. At end of wind vector, read ground distance from start of air plot to end of air plot position = 38 NM = Answer 3

5. If W/V is unknown and a fix is established at the end of the air plot, a line drawn on the navigation chart from the end of the air plot to the fix represents the W/V for the time of the air plot. This must be converted equivalent to a one hour wind

6. If a decision is made to do a 180° turn, the track required to the start of the air plot position can be found by rotating the compass rose to place the end of the wind vector on the centre line *above* the grommet and reading off the track direction under the TH index, TR 123° True = Answer 4

7. Drift is 8° right when outbound and drift on return will be 8° left. Therefore, double the drift and add to the TR 123° + 16° = TH 131° = Answer 5

Note: Heading vectors are marked *downwards*.
    Wind direction is marked *upwards*.

10 – Pressure Pattern Navigation

Introduction

Pressure Pattern Navigation was developed for long range navigation during the 1920–30s and was used originally by Zeppelin airships. It continued to be used during the 2nd World War through to the 1960's, being more suitable for relatively slower World War II aircraft and the piston-engine airliners that followed. It has been used by military jet aircraft in the intervening years.

Dr. John Carey Bellamy, a Civil Engineer in the USA, further developed the Pressure Pattern Navigation method and was credited with the Bellamy Drift Formula in 1943. The method is based on the use of constant-pressure upper-air charts where the isobars patterns provide the required information for flight-planning long range flights. The purpose is it find following winds, and the best ground-speed over longer routes than a Great Circle route can provide, and to flight-plan for single-heading flight and to calculate position lines and drift during flight. During flight, a radar altimeter is required to find the absolute altitude in conjunction with the pressure altimeter, to find the cross-wind component and drift, etc.

Pressure Pattern Navigation is not a pilot exam requirement, but is included in the Flight Navigator's syllabus.

Definitions

Zn	= Beam wind component in nautical miles
Vn	= Beam wind component in Knots
Vp	= Head- or tail-wind component in Knots
D1	= Height difference in feet between the absolute altitude and pressure altitude at position 'A'
D2	= Ditto at position 'B'
Dp	= The difference in feet between the readings of D1 and D2
K factor	= Wind factor for latitude. Use 'K' factor from the table opposite, for E6-B's with no latitude scale.

Note: To avoid confusion, if the aircraft is flying down the pressure gradient, drift will be to the right in the Northern Hemisphere and to the left in the Southern Hemisphere. Pressure Pattern Navigation cannot be used within 15° latitude of the Equator. For this reason, the Latitude scale runs from 15° to 90°.

Latitude 'K' Factor Table

Latitude	'K'	Latitude	'K'
15	83	36	37
16	78	37	36
17	73	38	35
18	69	39	34
19	66	40	33
20	63	41	33
21	60	42	32
22	57	43	31
23	55	44	31
24	53	45	30
25	51	46	30
26	49	47	29
27	47	48	29
28	46	49–51	28
29	44	52–54	27
30	43	55–57	26
31	42	58–61	24
32	41	62–66	24
33	39	67–72	23
34	38	73–87	22
35	37	88–90	21

Note: Use this table if your E6-B has no Latitude scale. This table is calculated from the formula:

95 – Zn, Lateral Drift

A position line can be drawn parallel to the aircraft's heading at a distance in nautical miles equal to the Zn Lateral Drift. Zn can be found given the TAS, pressure difference (Dp) in feet, and the mean latitude. The answer, Zn in nautical miles, can be read on the outer scale over the Latitude scale or from the 'K' factor on the inner scale. If Dp has a positive sign, the drift is to the left of heading; if the sign is negative, drift is to the right in the Northern Hemisphere. The rule is revered in the Southern Hemisphere.

Lateral drift (LD) = cross-wind component times flight endurance.

Formula:

$$\frac{Dp}{TAS} \cdots \frac{Zn\,(NM)}{'K'\,Latitude} = Answer$$

Given:
Dp = 115 feet
TAS = 450 Kts
Latitude 40° = 33 'K' factor

$$\frac{115\,Dp}{450\,Kts} \cdots \frac{8.55\,NM}{40°\,Latitude\,(33K)} = Answer$$

Note: Zn and distance gone in nautical miles can now be used to find the Drift Angle. See formula Drift Angle From Zn. Dp, TAS and latitude are also used to find Vp (Ground Speed).

96 – Drift Angle from Zn

This formula uses the beam wind component in nautical miles and the distance covered in nautical miles to find the Drift Angle. Use Formula 'A' for the trigonometry method and the formula 'B' for the wind slide method.

Formula 'A':

$$\frac{Zn}{Distance} \ldots \frac{Sine\ Drift\ Angle}{10} = Answer$$

Given:
Zn = 26.8 NM
Distance = 275 NM

$$\frac{26.8\ NM}{275\ NM} \ldots \frac{0.0974\ (5.6°)}{10} = Answer$$

Note: Zn must first be found using Formula 95, Zn Lateral Drift. With TAS, true heading, drift angle, ground speed, time and distance gone, the drift angle and therefore track made good (TMG), can be found.

Wind slide method: The Wind Slide Computer can also be used to solve Zn.

Formula 'B':

1. Select a Cardinal point (N, S, E or W)

2. Draw Zn distance above grommet 26.8 NM

3. Rotate compass rose to next cardinal point (90° either way)

4. Set slide to distance gone 275 NM

5. Read off Drift Angle at end of Zn distance = 5.6° = Answer

97 – Drift Angle from Vn

This formula is similar to the previous one but instead makes use of the beam-wind component in Knots (Geostrophic wind) and true air speed in Knots to find the sine drift angle.

Formula:

$$\frac{Vn}{TAS} \cdots \frac{\text{Sine Drift Angle}}{10} = \text{Answer}$$

Given:
Vn   = 34 Kts
TAS   = 235 Kts

$$\frac{34 \text{ Kts}}{235 \text{ Kts}} \cdots \frac{0.1446 \ (8.3°)}{10} = \text{Answer}$$

Wind slide method: The Wind Slide Computer can also be used to solve for Vn.

Method:

1. Select a Cardinal point (N, S, E or W)

2. Draw Vn distance above grommet 34 NM

3. Rotate compass rose to next cardinal point (90° either way)

4. Set slide to distance gone 235 NM

5. Read off Drift Angle at end of Zn distance 8.3° = Answer

98 – Modified Drift Formula

The Modified Drift Formula is also known as the Bellamy Drift formula after Dr. John Carey Bellamy, an American Civil Engineer, who devised the formula. This formula is similar to the Zn formula using the 'K' factor or Latitude scale, Dp, and TAS and trip time converted to decimal time, to find an average drift in degrees. The formula holds well with no great difference between the true air speed and ground speed.

Formula:

$$\frac{60 \times \text{'K'} \times \text{Dp}}{\text{TAS}^2 \times \text{Decimal time}} = \text{Drift Angle}° = \text{Answer}$$

Given:
Latitude 34° = 38K
TAS   = 315 Kts
Time   = 01.15 = 1.25 decimal Time
Dp   = –210

$$\frac{60 \times 38\text{K} \times 210 \text{ Dp}}{315 \text{ Kts} \times 315 \text{ Kts} \times 1.25} = 3.8° \ (4°) = \text{Answer}$$

99 – Geostrophic Cross-wind Component NAV

This formula gives the effective geostrophic cross-wind component and is independent of the aircraft's heading relative to the pressure gradient. The aircraft will drift perpendicularly to the pressure gradient at a rate depending on the geostrophic wind speed.

Formula:

$$\frac{Dp}{Distance} \dots \frac{Cross\text{-}wind\ Component}{K,\ or\ Latitude\ Scale} = Answer$$

Given:

Dp	= –525 feet
Distance	= 640 NM
Latitude 42°	= 32 K

$$\frac{525\ Feet}{640\ NM} \dots \frac{26\ Kts}{32\ K} = Answer$$

100 – Ground Speed Vp

With Pressure Pattern Navigation, we can find the average ground speed in Knots. Given the Latitude, TAS and Dp (which is also used to find Zn) the head-wind or tail-wind component can be found on the E6-B. The Vp component is either subtracted for a head-wind or added for a tail-wind to the TAS to find the ground speed.

Formula:

$$\frac{TAS}{K\,Latitude} \ldots \frac{Vp}{Dp} = Answer$$

Given:
TAS = 236 Kts
Dp = 170 feet
Latitude 49° = 28 K
(Assume a head-wind component).

$$\frac{236\,Kts}{28\,K,\,or\,Latitude\,49°} \ldots \frac{14.2\,Kts}{170\,Dp} = Answer$$

Therefore, TAS = 236 Kts minus Vp 14 Kts = ground speed 222 Kts = Answer

11 – Navigation by Latitude & Longitude

Introduction

Some navigation problems may call for calculating the track and the distance in nautical miles given the latitude and longitude of two locations, or vice versa. The parallels of latitude are equally spaced around the world where 1° latitude is equal to sixty nautical miles and one minute of latitude equals one nautical mile. To calculate distance in nautical miles or degrees and minutes for changes in latitude is quite simple. However, the lines of longitude are not parallel – except at the Equator – they converge to meet at each pole. Their distance apart varies with the Cosine of the latitude, known as Departure, which is measured in nautical miles. A basic understanding of trigonometry is required using the trig tables of Sine, Cosine, Tangent and Secant. The following abbreviations are used in the following formulas:

D.Lat   = Difference of latitude

D.Long = Difference of longitude

Dep   = Departure. Difference of longitude measured NM. Divide NM by 60 (degrees and minutes)

Cos Lat = Cosine of the latitude

101 – Latitude from Departure & D.Long

Given the departure in nautical miles and the difference of longitude (D.Long) in minutes of arc, the latitude can be found. The second and third formulas presented here are variations on this first formula.

Formula:

$$\frac{\text{Departure NM}}{\text{D. Long (Minutes)}} \cdots \frac{\text{Cos Lat}}{10} = \text{Answer}$$

Given:
Departure = 179 NM
D.Long = 185 minutes

$$\frac{179 \text{ NM}}{185 \text{ Minutes}} \cdots \frac{0.9675}{10} = \text{Answer}$$

Note: From tables Cos Lat = 0.9675 = Answer

102 – Distance from D.Long & Latitude

The distance in nautical miles (NM) between two positions can be found by using the departure and the cosine of the latitude.

Formula:

$$\frac{\text{Cos Lat}}{10} \cdots \frac{\text{Distance NM}}{\text{Departure}} = \text{Answer}$$

Given:
Cos Lat 43° = 0.7316
Departure = 1522 NM

$$\frac{0.7314}{10} \cdots \frac{1113 \text{ NM}}{1522 \text{ NM}} = \text{Answer}$$

103 – D.Long from Departure & Cos Lat

Transposing the first formula (Latitude from Departure & D.Long) the difference of longitude (D.Long) can be found given the departure and Cos Lat.

Formula:

$$\frac{\text{Departure}}{\text{Cos Lat}} \dots \frac{\text{D. Long}}{10} = \text{Answer}$$

Given:
Departure = 238 NM
Cos Lat 42° = 0.7431

$$\frac{238 \text{ NM}}{0.7431} \dots \frac{320}{10} = \text{Answer}$$

Therefore, D.Long = 5° 20' = Answer

104 – Track Required from Departure & D.Lat

The track required from the departure point to the destination can be found given the difference of latitude (D.Lat) and departure in NM using the formula Departure/D.Lat = Tan track angle. This formula is good for distances up to about 600 NM. For greater distances, spherical trigonometry is required to solve the problem.

Formula:

$$\frac{\text{Departure}}{\text{D. Lat}} \dots \frac{\text{Tan track angle}}{10} = \text{Answer}$$

Given:
Departure = 310 NM
D.Lat = 3° 06′ = 186 minutes

$$\frac{310 \text{ NM}}{186 \text{ Minutes}} \dots \frac{1.6667}{10} = \text{Answer}$$

From tables 1.6667 = 59° = Answer

The track angle found by trigonometry must now be converted to a direction relative to North (000 or 360°). Determine in which compass quadrant the track lies in.

Given:
Track angle = 059°

D.Lat	Departure	Quadrant	True Track
N	E	N-E	*000 + 59 = 059°*
N	W	N-W	*360 – 59 = 301°*
S	E	S-E	*180 – 59 = 121°*
S	W	S-W	*180 + 59 = 239°*

105 – Track Distance (N–S)

Having found the track required from the previous formula, the track distance can be found with this formula, given the D.Lat and Secant of the required track. Use D.Lat times Secant Track angle = Track Distance.

Formula:

$$\frac{D.\,Lat}{10} \cdots \frac{Track\ Distance}{Secant\ Track\ Angle} = Answer$$

Given:

D.Lat = 186 Minutes

Track Angle = 059°

$$\frac{286\ Minutes}{10} \cdots \frac{361\ NM}{1.9416} = Answer$$

106 – Change of Latitude

Given the Track Distance and Cos Track Angle, the change of Latitude can be found. Remember, 60 minutes (') = 1 degree (°)

Formula:

$$\frac{\text{Track Distance}}{10} \cdots \frac{\text{Change latitude}}{\text{Cos Track Angle}} = \text{Answer}$$

Given:
Track distance   = 183 NM
Cos Track Angle 56° = 0.5592

$$\frac{183 \text{ NM}}{10} \cdots \frac{102 \text{ Minutes}}{0.5592} = \text{Answer}$$

Therefore 102 minutes = 1° 42' = Answer

107 – Change of Longitude

Having found the change of latitude from the previous formula, the change of longitude (Ch. Long) can now be found. Given the track distance and Sine track angle, the third side of the triangle can now be found – the departure (NM), which must then be converted to Change of Longitude in minutes using the appropriate formula.

Formula A:

$$\frac{\text{Track Distance NM}}{10} \cdots \frac{\text{Departure NM}}{\text{Sin Track Angle}} = \text{Answer}$$

Given figures from the previous formula:
Track Distance = 183 NM
Sine Track Angle = 56° = 0.8290
Cos. Lat 54° 12' = 0.5850

$$\frac{183 \text{ NM}}{10} \cdots \frac{152 \text{ NM}}{0.8290} = \text{Answer}$$

Convert departure 152 NM to Change of Longitude in minutes of arc using Formula B:

Formula B:

$$\frac{\text{Dep NM}}{10} \cdots \frac{\text{D. Long}}{\text{Sin TR Angle}} = \text{Answer}$$

$$\frac{152 \text{ NM}}{10} \cdots \frac{260'}{0.8290} = \text{Answer}$$

Therefore, 260 minutes = 4° 20' Ch. Long = Answer

Note: This problem can be solved as one formula:

Formula C:

$$\frac{\text{Track Distance} \times \text{Sin Track Angle}}{\text{Cos Latitude}} = \text{Change Logitude} = \text{Answer}$$

$$\frac{183 \text{ NM} \times \text{Sin } 0.8290}{\text{Cos } 0.5850} = 260 \text{ Minutes} = 4° 20' = \text{Answer}$$

12 – Fuel Conversion Factors by Calculator

Introduction

The formulas presented in this section are better suited for solving on an electronic calculator. They have been included for those pilots studying the ALTP navigation syllabus where greater accuracy is required when calculating fuel conversion problems.

108 – Litres to Kg

Given:
275 Litres at Sp. G 0.71

Find weight in Kg:

Litres × Sp. G = Answer

275 Litres × 0.71 = 195 Kg = Answer

109 – Kilograms to Litres

Given:
540 Kg at Sp. G 0.74

Find quantity in Litres

$$\frac{y\ Kg}{Sp.\ G} = Answer$$

$$\frac{540\ Kg}{0.74} = 729.73\ Litres = Answer$$

110 – Litres to Pounds

Given:
875 Litres at Sp. G 0.70

Find weight in Pounds:

y Ltrs × 0.22 × Sp. G x 10 = Answer

875 × 0.22 × 0.70 × 10 = 1347.5 Lb = Answer

111 – Pounds to Litres

Given:
1665 Pounds at Sp. G 0.73

Find quantity in Litres:

$$\frac{y\ Lb \times 0.45}{Sp.\ G} = Answer$$

$$\frac{1665 \times 0.45}{0.73} = 1026.4\ Litres = Answer$$

112 – US Gallons to Pounds ATPL

Given:
1540 US Gallons at Sp. G 0.695

Find weight in Pounds:

$$\frac{y \text{ US Gallons} \times 5 \times \text{Sp.G} \times 10}{6} = \text{Answer}$$

$$\frac{1540 \times 5 \times 0.695 \times 10}{6} = 8919 \text{ Pounds} = \text{Answer}$$

113 – Pounds to US Gallons

Given:
8250 Lb at Sp. G 0.75

Find total US Gallons:

= y Lb × 5/6 × Sp. G x 10

= 8250 × 6/5 × 1/0.75 × 1/10

$$= \frac{8250 \times 1.2}{7.5}$$

= 1320 US Gallons = Answer

Note: 6/5 = 1.2

114 – US Gallons to Kg

Given:
10250 US G at Sp. G =0.71

Find weight in Kg

y US Gallons × 5/6 × 50/11 × Sp. G = Answer

10250 × 5/6 × 50/11 × 0.71 = Answer

10250 × 3.78 × 0.71 = 27508.95 Kg = Answer

Note: 5/6 × 50/11 = 3.78

115 – Kg to US Gallons

Given:

17630 Kg at Sp. G 0.74

Find quantity in US Gallons

y Kg × 6/5 × 11/50 × Sp. G = Answer

17630 × 0.264 × 1/0.74

$$\frac{17630 \times 0.264}{0.74} = 6289 \text{ US Gallons} = \text{Answer}$$

Note: 6/5 × 11/50 = 0.264.

116 – Imperial Gallons to Pounds **ALTP**

Given:
252 Imperial Gallons at Sp. G = 0.72
Find weight in pounds

Weight in pounds

y Imp G × Sp. G × 10

252 × 0.72 × 10

= 1814.4 Lbs = Answer

117 – Pounds to Imperial Gallons **ALTP**

Given:
2538 Lbs at Sp. G 0.74

Find quantity in Imperial Gallons

$$\frac{\text{y Lb}}{\text{Sp. G} \times 10}$$

$$\frac{2538}{0.74 \times 10}$$

= 343 Imperial gallons = Answer

13 – Miscellaneous Items

Introduction
In this last section we take a look at a few miscellaneous formulas, some of which may be helpful to the reader.

Convergence 'N' Factor Table   ATPL

The Convergence 'N' Factor table is taken from the trigonometry sine table and is suitable for quick mental calculations of Convergency and Conversion angles. However, it is only given to one decimal place and therefore, it obviously is not as accurate as a four-figure table.

Mean Latitude	'N' factor
0°	0.0
6°	0.1
8°	0.3
30°	0.5
45°	0.7
65°	0.9
90°	1.0

118 – Conversion Angle (Mercator)   ATPL

The Conversion Angle is the angle between the Great Circle track and Rhumb line bearings. When plotting radio bearings on a Mercator Chart, the radio bearing being a Great Circle line is curved and has to be converted to a straight line before plotting. This is done by applying the Conversion Angle (towards the Equator in either the north or south hemispheres).

D. Long (the difference of longitude) is the difference between the NDB beacon and the aircraft's position in degrees of longitude. The sine of the latitude is the average latitude between the NDB beacon and the aircraft's position. The following formula is used:

Formula:

$$\frac{½ \text{ D. Long}}{10} \dots \frac{\text{Conversion Angle}}{\text{Sine Latitude}} = \text{Answer}$$

Given:
½ D. Long = 22° = 11°
Sine Latitude = 54° = 0.7986

$$\frac{11°}{10} \dots \frac{8.8° \ (9°)}{0.7986} = \text{Answer}$$

119 – Convergency (Lambert)

Convergency is used on the Lamberts navigation charts when plotting radio bearings to allow for the angular difference of a Great Circle track measured at either the meridian of the aircraft or radio beacon. The formula requires the sine mean latitude and the difference of longitude (D. Long).

Formula:

$$\frac{D.\ Long}{10} \ldots \frac{Convergency}{Sine\ Latitude} = Answer$$

Given:
D. Long = 15°
Sine Latitude = 45 = 0.7071

$$\frac{15°}{10} \ldots \frac{10.6}{0.7071} = Answer$$

120 – Interpolation of Tables

Some aircraft performance and flight planning data are presented in tabular form, which require interpolation to extract values between given pairs of figures in the tables.

To find the difference between two known values given in the table, first place the difference on the E6-B outer scale over the '10' index. Then read the answer on the outer scale over the fractional difference, which is given as a decimal fraction. The answer is then added to the lower known value from the table.

Formula:

$$\frac{\text{Difference Between Values}}{10} \ldots \frac{\text{Intermediate Value}}{\text{Fractional Difference}} = \text{Answer}$$

Given:
Known values $\qquad$ = 600 and 1000
Difference $\qquad$ = 400
Fractional Difference $\quad$ = 2/3 = 0.666

Formula:

$$\frac{400}{10} \ldots \frac{266}{0.666} = \text{Answer}$$

Therefore, the intermediate value = 600 + 266 = 866 = Answer

121 – Gyro Wander

The Direction Indicator (DI) should be checked against the magnetic compass about every fifteen minutes during flight to correct for 'apparent wander', caused by the Earth's rotation. The error increases closer towards the poles at the rate of 15° sine latitude/hour. The sign is negative (−) in the Northern hemisphere and positive (+) in the Southern Hemisphere.

Formula:

$$\frac{\text{Laititude}}{10} \ldots \frac{\text{Drift Rate}}{\text{Sine Latitude}} = \text{Answer}$$

Given:
Sine Latitude = 32° South = 0.53

$$\frac{15° \text{ S}}{10} \ldots \frac{+8°/\text{hour}}{0.53} = \text{Answer}$$

Velocity

MPH or SM × 1.6093	= KMH or Km × 0.6214	= MPH or SM
MPH or SM × 0.8684	= Kts or NM × 1.1507	= MPH or SM
Kts × 1.8520	= KMH × 0.5397	= Kts
Kts × 1.6889	= FPS × 0.5924	= Kts

Length

NM × 1.8520	= Km × 0.5397	= NM
SM × 0.8690	= NM × 1.1507	= SM
Feet × 0.3048	= Metres × 3.2810	= Feet
Km × 0.6214	= SM × 1.6093	= Km
Metres × 1.0936	= Yards × 0.9144	= Metres
Metres × 3.2810	= Feet × 0.3048	= Metres

Mass

Kg × 2.2046	= Lb × 0.4536	= Kg

Fluid volume

Litres × 0.2642	= US Gals × 3.785	= Litres
Imp Gals × 1.2010	= US Gals × 0.8326	= Imp Gals
Imp Gals × 4.5460	= Litres × 0.2204	= Imp Gals

Temperature

°C to °F	$= (y\ °C × 1.8) + 32$	= °F
°F to °C	$= (y\ °F – 32) × 0.55$	= °C

Note: A temperature conversion scale is located on the E6-B slide rule face for quick conversions between Celsius and Fahrenheit. The above conversions are more accurate than the conversion scale.

Trigonometry Table

Sin		Cos		Tan	
0	0.0000	90		0	0.0000
5	0.0872	85		5	0.0875
10	0.1736	80		10	0.1763
15	0.2588	75		15	0.2679
20	0.3420	70		20	0.3640
25	0.4226	65		25	0.4663
30	0.5000	60		30	0.5773
35	0.5736	55		35	0.7002
40	0.6428	50		40	0.8391
45	0.7071	45		45	1.0000
50	0.7660	40		50	1.1918
55	0.8191	35		55	1.4281
60	0.8660	30		60	1.7321
65	0.9063	25		65	2.1445
70	0.9397	20		70	2.7475
75	0.9659	15		75	3.7321
80	0.9848	10		80	5.6713
85	0.9962	5		85	11.4300
90	1.0000	0		90	–
Sin		**Cos**			

Glossary of Terms

Absolute altitude The altitude above the terrain as indicated by a radar or radio altimeter.

Ambient Refers to meteorology condition surrounding the aircraft

'Arm' The distance measured from the datum of a location within the aircraft.

10 Arrow index The '10 index' is located on the circular slide rule's inner scale. This index is frequently used for various calculations and is therefore marked with a long thin arrow.

60 Rate Index arrow A frequently used index mark is the '60' index, marked by a triangular arrow head. Commonly used with time, speed and distance calculations.

Air Speed Window Used for air speed, and density altitude calculations.

Air vector (AV) When the TH & TAS are plotted on the wind slide entre line it is referred to as the air vector

Altitude window Used for altitude calculations.

AUW The present weight of the aircraft.

Calculated air temperature (COAT) The temperature as indicated on the outside air temperature gauge after correcting for aerodynamic heating (temperature rise) due to increasing true air speed.

Calibrated air speed Calibrated air speed (CAS) is the air speed indicator reading corrected for instrument and position error. It is also known as Rectified air speed.

Centre line as TH & TAS The wind-cross is drawn above the grommet. The wind direction is from the TH & TAS vector on the drift line and speed arc, to the TR & G/S on the centre line.

Centre line as TR & G/S The wind cross is placed below the grommet. The wind vector acts from the TH & TAS on the centre line to the TR and G/S on the drift line and speed arc.

CG The aircraft's centre of gravity. The point of balance must lie within the CG limits.

CG limits The CG fore and aft limits define the allowable range within which the aircraft's centre of gravity must lie. Flight outside of the limits will compromise aircraft stability and control.

COAT The calculated outside air temperature, corrected for dynamic heating (skin friction).

Compass rose The compass rose is located around the edge of the circular disc on the wind slide numbered with compass headings from zero to 359 degrees.

Compass rose fixed scale On the fixed scale of the compass rose is the true Heading Index and either side is the Port and Starboard drift scale combined with the Variation east and west scale. It is marked from zero degrees to about 50° either side of the Heading Index.

Compressibility correction Compressibility becomes evident at speeds above 200 Knots and 10 000 feet altitude; it affects the air speed indicator readings in the form of over-reading. It requires conversion from CAS to EAS.

Convergency The angle between the great circle bearing and Rhumb line bearing to correct the difference when plotting on a Lambert's chart.

Conversion angle The angle between the great circle rack and the Rhumb line bearing. It is used to correct radio nav bearings when plotting on a Mercator chart.

Corrected outside air temperature (COAT) Due to kinetic heating with increasing air speed, the temperature over reads, the amount deducted depending on air speed.

Critical point Position on route from A to B where it is as quick to return to departure point as it is to continue to destination. Also known as the PNR or Equi-time point (ETP).

Cross-wind component The wind acting at right angles to the runway. See head-wind component.

Cursor A rotating cursor with an index line is a helpful tool on the circular slide rule to facilitate reading adjacent scales.

D.Lat Refers to the difference between two latitudes measured in degrees, minutes and seconds of arc.

D.Long Refers to the difference of longitude between two places in degrees, minutes and seconds of arc.

Density altitude Pressure altitude adjusted for non-standard temperature determines the density altitude, which is the height in feet above sea level measured in terms of air density. The combination of pressure altitude and temperature determines the density altitude.

Departure The distance between two longitudes measure in degrees and minutes of arc converted to nautical miles.

Distance gone The distance from the departure point to present position along the route.

Distance scales The outer scale on the circular slide rule contains the distance scale marked as Naut. M., Stat.M., or Km.

Lambert's chart 1:500 000 sectional charts are usually of the Lambert's projection where longitudes converge to the nearest pole.

Latitude scale The latitude scale is located on the circular slide rule and is aligned with the 'K' factor on the inner scale. Used when calculating Pressure Pattern Navigation problems. Use either latitude scale or 'K' factor from table.

LEMAC Leading edge (of) mean aerodynamic chord.

Low & high speed wind slide On either side of the wind slide are the scales from approximately 40 KTAS to 300 KTAS on the low speed slide, and from approximately 150 KTAS to 1000 KTAS on the high speed slide.

Mach number Mach number is a dimensionless number defined as the ratio of the aircraft's true air speed to the speed of sound in the surrounding air. The Mach number varies as the square root of the absolute temperature.

MAUW Maximum allowable weight of the aircraft.

Mercator's chart Distinctive by its lines of latitude being parallel and also parallel lines of longitude. Scale increases by the secant of the latitude towards the poles. Used mainly for long range navigation.

Nautical miles (NM) One NM = 6080 feet, the most commonly used measurement for navigation.

Nautical miles index Used for conversions to or from statute miles or kilometres.

NDB Non directional beacon, a radio nav aid.

Outer scale Logarithm scale. Located on the circular slide rule and also known as the distance or TAS scale. Several important conversion scales lie on the outer edge of the 'outer scale', e.g., Specific Gravity, Distance conversions, etc. It is also known as the 'A' scale.

Percentage (%) MAC The mean aerodynamic chord. Used for CG position calculations in relation to the aircraft's mean aerodynamic chord.

PNR The point of no return. Also known as the critical point.

Point of safe Diversion The position en route where an aircraft can divert to an alternate airfield with Safe Endurance.

Pressure altitude The altitude indicated by the altimeter when set to QFE 1013 hPa (29.92 inches).

Pressure Pattern Navigation An old system of long range navigation introduced in the 1930 – 40's based on contour patterns of constant pressure, upper air meteorology charts. Uses the difference between absolute and pressure altitudes for calculating single-heading flight, etc

Radius of Action The furthest distance in nautical miles a plane can fly away from its base and return within its safe endurance. See safe Endurance.

Required track The direct route required to fly from A to B.

Rhumb line These are constant headings but cover greater distance than great circle tracks.

RoC The rate of climb is measured in feet per minute.

RoD The rate of descent is measured in feet per minute.

Safe endurance Determined by fuel available and consumption measured in time. Reserve fuel is excluded.

Sp. G The specific gravity of fuels and oils, is the weight of fuel relative to water at +4° C.

Speed arcs The curved arc lines represent the true air speed or ground speed on the wind slide plotting disc.

Speed of closing The relative speed at which two aircraft are intercepting each other.

Square graph A square graph is located at the bottom of the low-speed wind slide, useful for working some of the problems involved in navigation, e.g. cross-wind and head-wind components affecting take-off and landing, and for calculating en route diversions, winds on the climb, etc.

Tail-wind A tail-wind is an advantage for increased ground speed in flight. Adverse affect during take-off or landing.

Temperature conversion scale A reference scale for conversions between °F and °C. Found on the lower edge of the circular slide rule.

Temperature rise scale The required correction to the indicated air temperature due to increasing true air speed, can be found from this scale.

ToC The position when climb is completed on reaching the initial cruise altitude.

ToD The top of descent position at the end of the en route cruise segment.

Track The path the aircraft is following, or is intended to follow, over the Earth's surface.

Track made good The true path over the ground, which the aircraft has covered. Not necessary the required track but preferable when they coincide.

True heading The direction the aircraft is pointing in relation to true north.

VOR Acronym for Very high frequency, Omni directional, Radio range, a radio navigation aid.

Weighted wind components Averaged en route, or climb, wind velocities factored for time in each zone.

Wind arm Some E6-B's have a rotatable arm attached to the grommet, or centre of the plotting disc to assist in wind triangle problems. Ideal for simple wind problems, but more of a hindrance for advanced wind finding methods.

Wind cross A pencil cross (or dot) drawn on the E6-B wind slide to represent the end of the W/V vector.

Wind drift correction scale Located on the wind slide's fixed portion adjacent to the wind circular rose. Used for wind drift, magnetic and compass headings calculations.

Wind lanes Observed when flying at relatively low levels over the sea. The wind makes visible patterns inline with the wind direction at sea-level. From these lines wind direction can be noted for drift calculations.

Wind slide grid The wind slide is the adjustable sliding plate behind the circular slide rule for plotting graphical nav problems. One side is graduated for low-speed flight (40 to 300 Knots) and the other side is graduated for high-speed flight (150 – 1000 Knots) or thereabouts. On most E6-B's, the low-speed slide also includes a square grid section.

Wind triangle A graphical method of using trigonometry to calculate W/V, TH, TR, G/S and TAS on the E6-B's wind slide.

W/V Represents the wind velocity vector for working wind triangle problems.

www.ingramcontent.com/pod-product-compliance
Lightning Source LLC
Chambersburg PA
CBHW020203200326
41521CB00005BA/232